Experiments in Electronics Fundamentals and Electric Circuits Fundamentals

Eighth Edition

David M. Buchla

Pearson

Harlow, England • London • New York • Boston • San Francisco • Toronto • Sydney
Dubai • Singapore • Hong Kong • Tokyo • Seoul • Taipei • New Delhi
Cape Town • São Paulo • Mexico City • Madrid • Amsterdam • Munich • Paris • Milan

Editor in Chief: Vernon Anthony
Acquisitions Editor: Wyatt Morris
Editorial Assistant: Chris Reed
Director of Marketing: David Gesell
Senior Marketing Assistant: Les Roberts
Senior Managing Editor: JoEllen Gohr
Project Manager: Rex Davidson
Senior Operations Supervisor: Pat Tonneman
Operations Specialist: Laura Weaver

Art Director: Candace Rowley
Cover Designer: Rachel Hirchi
Cover art: Jan Neville/iStockphoto
Lead Media Project Manager: Karen Bretz
Printer/Binder: Bind-Rite Graphics
Cover Printer: Lehigh-Phoenix Color
Text Font: Times Roman

LabVIEW, Multisim, NI, Ultiboard, and National Instruments are trademarks and trade names of National Instruments. Other product and company names are trademarks or trade names of their respective companies.

18 2021

ISBN 13: 978-0-13-506327-9
ISBN 10: 0-13-506327-2

Preface

This laboratory manual is designed to accompany *Electronics Fundamentals: Circuits, Devices, and Applications,* Eighth Edition, and *Electric Circuits Fundamentals,* Eighth Edition, both by Thomas L. Floyd and David M. Buchla. Revisions to the Application Assignment worksheets correspond to changes in these assignments in the text. The text/lab manual correlation is shown on page vii.

New in this edition is a motor project in Appendix A, which can be used in conjunction with the expanded coverage on dc motors in Chapter 7 of the text. The project requires much more time than is available in a typical lab period, so it is given as a supplement to the normal lab sequence. The parts for the motor are listed only with the project and are not included in the overall materials list in Appendix B. I appreciate the idea for the motor from Mike Halbern of Sierra College.

PowerPoint® slides are available to instructors for each experiment and include a number of new slides added for this edition. The slides are designed to review the experiment and include a "Trouble" and "Related Problem." These can be used for class discussion with the solution presented on a following slide or slides. The slides are available free of charge to instructors using this manual.

To access supplementary materials online, instructors need to request an instructor access code. Go to **www.pearsonhighered.com/irc**, where you can register for an instructor access code. Within 48 hours after registering, you will receive a confirming e-mail, including an instructor access code. Once you have received your code, go to the site and log on for full instructions on downloading the materials you wish to use.

Multisim® computer simulations have been updated to both version 9 and version 10 files and are available for 22 experiments. Experiments with Multisim files are marked with a computer icon in both the Table of Contents and at the top of the experiment. Multisim problems are not a part of the laboratory experimental work and can be eliminated with no effect on the experiments. Each experiment with a Multisim component has four files. Files are coded with the name EXPx-y-z. The letters x-y stand for a particular experiment number and figure number within the experiment. The letter z refers to either "nf" for "no fault" or f1, f2, or f3 for circuits with faults. The suffix depends on the Multisim version in use, so it is not shown. Multisim files are available at www.prenhall.com/floyd. Select the text and follow the links to the files. For help getting started with Multisim, see pages 1–6 of this manual or the appendix in the text, which includes downloading instructions for a trial version of Multisim and other resources.

The key elements of the manual are its workbook format and the flexibility it offers for various approaches to lab work.

- **Workbook format.** Data tables are located close to their first reference, simplifying the entering of data. Space is provided for observations within the experiment and for answering related questions. Space is also provided for a written conclusion. The workbook format continues with blank tables for the Application Assignments given in the text and with the Checkup exercises. Checkups include both lab and text information.
- **Flexibility.** At the end of each lab, the For Further Investigation section (with less structure than the experiment) can be assigned as part of the experiment or used as an extra credit assignment, depending on the particular time allowed and instructor preference. For Further Investigation is intended to go beyond the basic experiment. Flexibility is further enhanced by the inclusion of the Multisim computer troubleshooting exercises, the Checkup exercises, the Application Assignment (and the Related Experiment), and the PowerPoint® slides.

Each experiment contains the following parts:

- *Reading:* These reading assignments are referenced to *Electronics Fundamentals: Circuits, Devices, and Applications* and *Electric Circuits Fundamentals.*

- *Objectives:* This is a statement of what the student should be able to do after completing the experiment.
- *Materials Needed:* This lists the components and small items required to complete the experiment. It does not include the equipment found at a typical lab station.
- *Summary of Theory:* This reinforces the important concepts in the text by reviewing the main points prior to the laboratory experience. In most cases, specific practical information needed in the experiment is presented.
- *Procedure:* This section contains a relatively structured set of steps for performing the experiment. Needed tables, graphs, and figures are placed near their initial reference to avoid confusion. Laboratory techniques, such as operation of both analog and digital oscilloscopes, are given in detail.
- *Conclusion:* A space is provided for the student to write a conclusion to the experiment. A sample conclusion is given in Experiment 1.
- *Evaluation and Review Questions:* This section contains five or six questions that require the student to draw conclusions from the laboratory work and check his or her understanding of the concepts. Troubleshooting questions are frequently presented.
- *For Further Investigation:* This section contains specific suggestions for additional related laboratory work. A number of these lend themselves to a formal laboratory report, or they can be used as an enhancement to the experiment.

Following the experiments designed for a specific chapter of the texts are the Application Assignment and Checkup. These pages can be removed from the book. The Application Assignment begins with an answer page for the student to complete. The assignment is correlated to the application problem given at the end of the text chapters (beginning with Chapter 2). Most Application Assignments include a Related Experiment, which adds a problem that requires a laboratory solution. The Checkup begins with ten multiple-choice questions and includes questions and problems from the text and the laboratory work, with more emphasis on the laboratory work. These items are cross-indexed on page vii.

Each laboratory station should contain a dual-variable regulated power supply, a function generator, a multimeter, and a dual-channel oscilloscope (either an analog or digital type). It is useful if the laboratory is equipped to measure capacitors and inductors. In addition, a meter calibrator, a commercial Wheatstone bridge, and a transistor curve tracer are useful but not required. A list of all required components is given in Appendix B, with materials for the For Further Investigation and Related Experiment listed separately.

Although this manual is specifically designed to follow the sequence of *Electronics Fundamentals: Circuits, Devices, and Applications* and *Electric Circuits Fundamentals**, it can be used with other texts by ignoring the reading references. The experiments work equally well for schools using electron flow or conventional current flow, as no specific reference to either is made in the experiments.

I have enjoyed the close collaboration with Tom Floyd on this manual. I also would like to thank the reviewers who have given excellent suggestions for improving the experiments for this and previous editions. These are Jacob Baraseh, DeVry University; Carl A. Jensen; William P. Kist, New England Institute of Technology at Palm Beach; Harold Orner, Heald College; William P. Perclakes, Heald College; Carl F. Ervin, Texas State Technical Institute; George Borchers, Ernest Arney, and David Terrell of ITT Technical Institute; and Bill Frandrup at Yuba College. I would also like to thank Rex Davidson of Prentice Hall for his help and commitment to the project, Lois Porter for her superb copyediting work, and Mark Walters of National Instruments for suggestions on Multisim. Finally, I want to acknowledge the support and encouragement of my wife, Lorraine. She is the headlight at the end of the tunnel.

David Buchla

*Experiments 1-30 and the motor project in Appendix A apply to both books; Experiments 31-45 apply only to *Electronics Fundamentals: Circuits, Devices, and Applications*.

Contents

*Computer icons indicate Multisim files are available for this experiment.

Introduction to the Student

PREPARING FOR LABORATORY WORK

The purpose of experimental work is to help you gain a better understanding of the principles of electronics and to give you experience with instruments and methods used by technicians and electronic engineers. You should begin each experiment with a clear idea of the purpose of the experiment and the theory behind the experiment. Each experiment requires you to use electronic instruments to measure various quantities. The measured data are to be recorded, and you need to interpret the measurements and draw conclusions about your work. The ability to measure, interpret, and communicate results is basic to electronic work.

Preparation before coming to the laboratory is an important part of experimental work. You should prepare in advance for every experiment by reading the *Reading, Objectives,* and *Summary of Theory* sections before coming to class. The *Summary of Theory* is *not* intended to replace the theory presented in the text—it is meant only as a short review to jog your memory of key concepts and to provide some insight to the experiment. You should also look over the *Procedure* for the experiment. This prelab preparation will enable you to work efficiently in the laboratory and enhance the value of the laboratory time.

This laboratory manual is designed to help you measure and record data as efficiently as possible. Techniques for using instruments are described in many experiments. Data tables are prepared and properly labeled to facilitate recording. Plots are provided where necessary. You will need to interpret and discuss the results in the section titled *Conclusion* and answer the *Evaluation and Review Questions.* The *Conclusion* to an experiment is a concise statement of your key findings from the experiment. Be careful of generalizations that are not supported by the data. The conclusion should be a specific statement that includes important findings with a brief discussion of problems, or revisions, or suggestions you may have for improving the circuit. It should directly relate to the objectives of the experiment. For example, if the objective of the experiment is to use the concept of equivalent circuits to simplify series-parallel circuit analysis (as in Experiment 10), the conclusion can refer to the simplified circuit drawings and indicate that these circuits were used to compute the actual voltages and currents in the experiment. Then include a statement comparing the measured and computed results as evidence that the equivalent circuits you developed in the experiment were capable of simplifying the analysis.

THE LABORATORY NOTEBOOK

Your instructor may assign a formal laboratory report or a report may be assigned in the section titled *For Further Investigation.* A suggested format for formal reports is as follows:

1. *Title and date.*
2. *Purpose:* Give a statement of what you intend to determine as a result of the investigation.
3. *Equipment and materials:* Include a list of equipment model and serial numbers that can allow retracing if a defective or uncalibrated piece of equipment was used.
4. *Procedure:* Give a description of what you did and what measurements you made.
5. *Data:* Tabulate raw (unprocessed) data; data may be presented in graph form.
6. *Sample calculations:* Give the formulas that you applied to the raw data to transform them to processed data.
7. *Conclusion:* The conclusion is a specific statement supported by the experimental data. It should relate to the objectives for the experiment as described earlier. For example, if the purpose of the experiment is to determine the frequency response of a filter, the conclusion should describe the frequency response or contain a reference to an illustration of the response.

GRAPHING

A graph is a pictorial representation of data that enables you to see the effect of one variable on another. Graphs are widely used in experimental work to present information because they enable the reader to discern variations in magnitude, slope, and direction between two quantities. In this manual, you will graph data in many experiments. You should be aware of the following terms that are used with graphs:

abscissa: the horizontal or *x*-axis of a graph. Normally the independent variable is plotted along the abscissa.

dependent variable: a quantity that is influenced by changes in another quantity (the independent variable).

graph: a pictorial representation of a set of data constructed on a set of coordinates that are drawn at right angles to each other. The graph illustrates one variable's effect on another.

independent variable: the quantity that the experimenter can change.

ordinate: the vertical or *y*-axis of a graph. Normally the dependent variable is plotted along the ordinate.

scale: the value of each division along the *x*- or *y*- axis. In a linear scale, each division has equal weight. In a logarithmic scale, each division represents the same percentage change in the variable.

The following steps will guide you in preparing a graph:

1. Determine the type of scale that will be used. A linear scale is the most frequently used and will be discussed here. Choose a scale factor that enables all of the data to be plotted on the graph without being cramped. The most common scales are 1, 2, 5, or 10 units per division. Start both axes from zero unless the data covers less than half of the length of the coordinate.

2. Number the *major* divisions along each axis. Do not number each small division as it will make the graph appear cluttered. Each division must have equal weight. Note: The experimental data is *not* used to number the divisions.

3. Label each axis to indicate the quantity being measured and the measurement units. Usually, the measurement units are given in parentheses.

4. Plot the data points with a small dot with a small circle around each point. If additional sets of data are plotted, use other distinctive symbols (such as triangles) to identify each set.

5. Draw a smooth line that represents the data trend. It is normal practice to consider data points but to ignore minor variations due to experimental errors. (Exception: calibration curves and other discontinuous data are connected "dot-to-dot.")

6. Title the graph, indicating with the title what the graph represents. The completed graph should be self-explanatory.

SOLDERLESS BREADBOARDS

The solderless breadboard (also called "Experimenter Socket" or "protoboard") is a quick way to build circuits for test, so it is widely used in schools for building circuits such as those in this lab manual. Spring-loaded connectors are internal to the board and form groups of holes that are common connection points. There are different sizes and arrangements of solderless breadboards, but most are quite similar to the one described here (the Radio Shack #276–174).

Solderless breadboards are designed to connect small parts such as resistors, capacitors, or transistors together. Wires are also inserted in holes to connect them to parts or other wires. Wires should

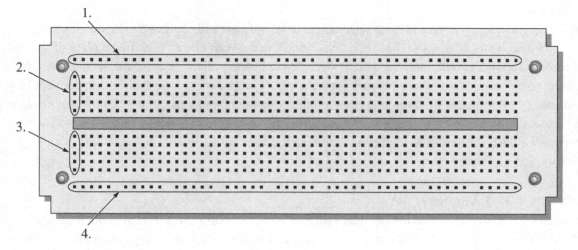

Figure I–1

be #22 or #24 solid core wire (not stranded) and be stripped about 3/8″. Figure I–1 shows a typical solderless breadboard. The following points are indicated on Figure I–1:

1. A horizontal row generally used by the experimenter for power or ground. These holes are all interconnected so that a wire or part inserted in any one is connected to all other holes in the row. Some boards have this row "broken" in the middle or have more than one long row.

2. A vertical column of five holes connected together internally with a spring-loaded connector. Any part inserted in any one of these holes can be joined to another part by inserting the second part into another of these holes. This group of five holes is isolated from all other groups of holes.

3. Another vertical column of five holes connected together internally. These are separate from the holes described in 2.

4. Another horizontal row with all of the holes connected together.

Figure I–2 shows an example of three resistors connected in series (one path) as you will do in Experiment 4. The choice of which holes to use is up to you but the ends of two resistors must be connected in holes that are joined in the board in order to connect them together. Other examples of wiring are given within the lab manual.

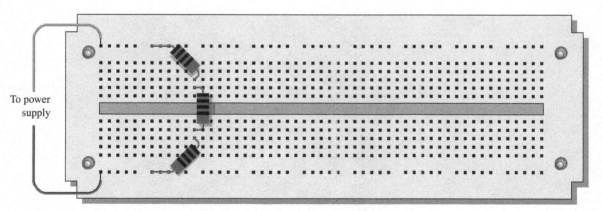

Figure I–2

WIRING HINTS

When wiring a circuit, you should keep wiring neat to be able to easily follow the layout. The best way to keep wiring neat is to cut wires to the size needed; don't make your circuit look like a plate of spaghetti. Don't force components with large leads into the holes in boards; this will cause the spring-loaded connectors to be permanently distorted and not hold other components properly. You can connect larger components by clipping an alligator lead between the component and a piece of #22 or #24 wire. Don't insert wires more than 1/4″ into the holes. When wires are inserted too far, they find a way to short another row or push the spring-loaded connectors out the back of the board. A "bad board" can be a troubleshooting nightmare.

SAFETY IN THE LABORATORY

The experiments in this lab book are designed for low voltages to minimize electric shock hazard; however, never assume that electric circuits are safe. A current of a few milliamps through the body can be lethal. In addition, electronic laboratories often contain other hazards such as chemicals and power tools. For your safety, you should review laboratory safety rules before beginning a course in electronics. In particular, you should

1. Avoid contact with *any* voltage source. Turn off power before you work on circuits.
2. Remove watches, jewelry, rings, and so forth before you work on circuits—even those circuits with low voltages—as burns can occur.
3. Know the location of the emergency power-off switch.
4. Never work alone in the laboratory.
5. Keep a neat work area and handle tools properly. Wear safety goggles or gloves when required.
6. Ensure that line cords are in good condition and grounding pins are not missing or bent. Do not defeat the three-wire ground system in order to make "floating" measurements.
7. Check that transformers and instruments that are plugged into utility lines are properly fused and have no exposed wiring. If you are not certain about procedure, check with your instructor first.
8. Report any unsafe condition to your instructor.
9. Be aware of and follow laboratory rules.

Circuit Simulation and Prototyping
Using Multisim and NI ELVIS

DESIGN, SIMULATION, PROTOTYPING, AND LAYOUT

Electronics has changed rapidly and has become a part of an increasing array of products. As circuits and systems become more advanced, circuit designers rely on computers assist in the design process. The computer has become a vital part of this process. Producing a new circuit for a product can be broken into four main steps: design, simulation, prototyping, and layout. The last three steps usually involve a computer.

A circuit designer begins the development process with an idea for solving a problem. The idea is developed into a circuit, and the designer usually calculates the expected results by hand to get a good idea of circuit behavior. The second step is to enter the schematic into a computer (this is called "schematic capture") and test it with a circuit simulation program such as *Multisim*. The third step is to construct and test a prototype, which may reveal a hidden or unforeseen problem. Prototyping can also be computer aided as will be discussed. Sometimes the simulation and prototyping steps are repeated to refine the design. Finally, the design is ready to be implemented. This fourth step is done by transferring the circuit design to a printed circuit board (PCB) using a graphical layout tool like *Ultiboard* to determine the optimum placement and interconnection of the components.

The focus here is for two of the steps in the design process —simulation and prototyping. As a design aid, Multisim is one of the most widely used circuit design and simulation tools in industry. Multisim is also used in educational environments because it can simulate circuits quickly, including circuits with troubles. Many Multisim problems have been prepared for this lab manual and the text to introduce you to computer design and simulation tools and to give you troubleshooting practice in a simulated circuit. Although computer simulations are useful and allow you to test parameters that may be difficult, unsafe, or impossible to attain in the lab, they should not be considered a replacement for careful lab work.

The traditional circuit prototype in electronics classes is usually constructed on a solderless protoboard ("breadboard"). The prototype circuit can be tested with stand-alone instruments in the lab. Alternatively, the circuit can be tested with a complete prototyping system like National Instrument's Educational Laboratory and Virtual Instrumentation Suite (NI ELVIS). The NI ELVIS system has over 12 built-in instruments and an interface that can communicate with Multisim. Multisim can also simulate the NI ELVIS interface, as will be shown later.

MORE ABOUT MULTISIM

Multisim software integrates powerful SPICE simulation and schematic capture into a highly intuitive electronics lab on the computer. Use the Multisim Student Edition to enhance learning by:

- Building circuits in a simple, easy-to-learn environment
- Simulating and analyzing circuits for homework and prelab assignments
- Breadboarding in 3D at home before lab sessions
- Creating designs of up to 50 components using a library of nearly 4,000 devices
- Integrating with the NI ELVIS prototyping workstation

While using industry-standard SPICE in the background, you can take advantage of the Multisim drag-and-drop interface to make circuit drawing, wiring, and analysis simple and easy to use. You can

1

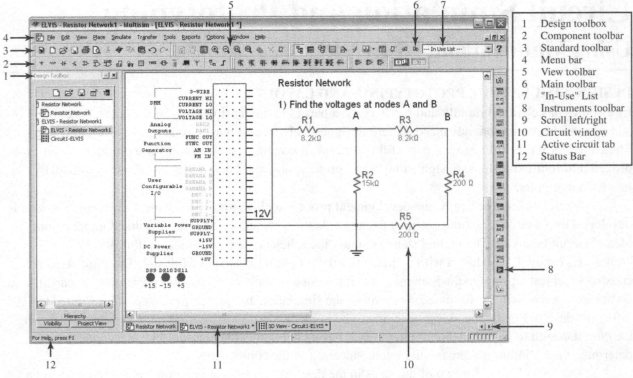

Figure I–3 The Multisim environment

1	Design toolbox
2	Component toolbar
3	Standard toolbar
4	Menu bar
5	View toolbar
6	Main toolbar
7	"In-Use" List
8	Instruments toolbar
9	Scroll left/right
10	Circuit window
11	Active circuit tab
12	Status Bar

build circuits from scratch and learn about design performances using built-in virtual instruments and probes in an ideal environment for experimentation. With the 3D breadboard, you can easily take the leap from circuit diagram to real-world implementation.

Create an example from the manual to get familiar with the Multisim environment. First, launch Multisim and open a new schematic window (**File» New» Schematic Capture**), such as the one shown in Figure I-3. This figure outlines the various elements of the Multisim environment. Design circuits in the circuit window by placing components from the component toolbar. Clicking on the component toolbar opens the component browser. Choose the family of components and select an individual component to place on the circuit window by double-clicking on it.

Once you have selected a component, it attaches itself and "ghosts" the mouse cursor. Place the component by clicking again on the desired location in the schematic. If you are new to Multisim, you should use the **BASIC_VIRTUAL** family of components, which you can assign any value.

The next step is to wire components together. Simply left-click on the source terminal and then left-click on the destination terminal. Multisim will automatically choose the best path for the virtual wire between the two terminals. Always be sure your circuit has a source and a ground or reference, which are components found in the sources group. Once the circuit has been fully captured, it can be simulated. Results of the simulation can then be used as a comparison with the physical circuit.

As an example of a simple Multisim circuit, a resistor network is shown in Figure I-3. The network is connected to a virtual interface that represents inputs and outputs on the NI ELVIS station. To create this type of simulated circuit, first create a new NI ELVIS schematic by clicking **File » New » ELVIS II Schematic**. Place virtual resistors by clicking on the resistor icon of the components toolbar. Change the values of the resistors to match Figure I-3. Wire the NI ELVIS positive supply and ground reference to your circuit, and the circuit is ready to simulate.

Alternatively, you can create a traditional schematic by clicking **File » New » Schematic Capture**. If you do this, you will need to manually place power and ground components. This is useful if you are using traditional stand-alone instruments to analyze your circuit. Once you have drawn the circuit, there are many ways to analyze your design using the simulation software.

Assume you want to know the steady-state voltages of points A and B in the circuit (V_A and V_B). The voltages can be calculated by methods you will learn later in Chapter 6 of the text. For now, we'll assume you have applied these methods to obtain the prediction that $V_A = 4.81$ V and $V_B = 0.245$ V.

MORE ABOUT NI ELVIS

NI ELVIS is a LabVIEW-based design and prototyping environment based on LabVIEW for university science and engineering laboratories. NI ELVIS consists of virtual instruments (VIs), or functions, based on LabVIEW; a multifunction data acquisition device; and a custom-designed benchtop workstation and prototyping board. This combination provides a ready-to-use suite of instruments for educational laboratories. Figure I-4 shows the NI ELVIS workstation.

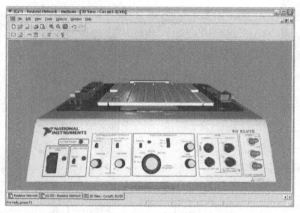

Figure I–4   NI ELVIS workstation

After a circuit has been simulated successfully and you are satisfied with its operation, you can build a prototype. Prototyping hardware and instruments are designed to allow you to quickly construct circuits and test them in the real world. Prototyping is valuable for verifying the operation of the design and for uncovering real-world inconsistencies that simulation was not able to predict. For example, the simulation might not show the effect of unexpected spikes from the wall power or interference from a cell phone. It is critically important that there is a high correlation between real-world measurements and simulation results. A close match assures you that the design will effectively solve the problem.

Your lab may be equipped with separate instruments or you may have them provided as a functional package, such as the NI ELVIS prototyping hardware system. The advantage of the NI ELVIS system is that it can aid in the prototyping process by allowing you to construct circuits first on a "virtual protoboard". After constructing the virtual circuit, you can transfer it to the hardware in the NI ELVIS system. The ideal electronics laboratory will provide an easy way for comparing measurements to simulations. The NI ELVIS system integrates naturally with Multisim to provide a seamless transition from design to prototyping. All the measured data is available in a single environment that allows for quick and easy comparison of simulations to real-world measurements.

ANALYZING THE RESISTOR NETWORK USING A REAL-WORLD PROTOTYPE

Returning to the resistor network introduced with Multisim as an example, you can continue to the prototyping step. You can build the circuit on a standard protoboard and test it with the laboratory instruments to complete the design.

Alternatively, you can repeat the simulation step but this time as a prototype using the 3D Virtual NI ELVIS from within Multisim. 3D prototyping allows you to learn more about using breadboards and experiment with your designs in a risk-free environment. To construct a 3D ELVIS prototype, open the 3D breadboard by clicking **Tools » Show Breadboard**. Place components and wires to build up your circuit. The corresponding connection points and symbols on the NI ELVIS schematic will turn green, indicating the 3D connections are correct. If you created a traditional schematic, you will see a standard breadboard. Figure I-5 shows the virtual NI ELVIS (background image) and the actual hardware foreground image). Figure I-6 shows the circuit on NI ELVIS prototyped in Multisim and ready for construction. Once the layout of your circuit has been verified using the 3D virtual environment, it can be physically built on the NI ELVIS.

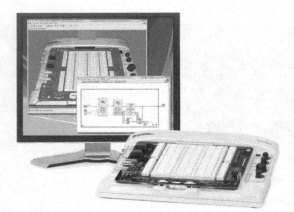

Figure I–5 Virtual NI ELVIS (background) and in hardware (foreground)

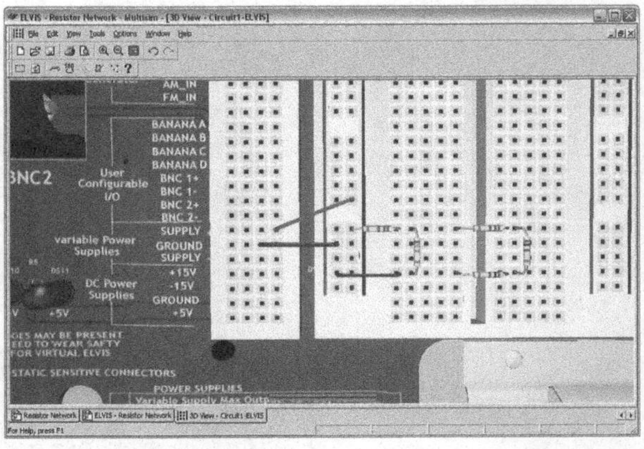

Figure I–6 Resistor network prototyped in Multisim

Figure I-7 shows the resistor network on an NI ELVIS protoboard. The variable power supply is connected to the input of the resistor network. Connections to other instruments such as the oscilloscope, DMM, AM and FM modulator lines are also available on the breadboard for more advanced applications.

After you have wired the circuit, you can use the variable power supply provided by NI ELVIS to supply 12 volts (V_S) as needed by the experimental circuit and measure the voltages at A and B using interactive graphical software.

Figure I-8 shows the result of SignalExpress, an interactive measurement tool based on LabVIEW. SignalExpress provides a step-by-step interface to a computer to allow you to perform measurements. The screen shows the measured voltage on the prototype circuit displayed on the computer. If desired, it can be exported to an Excel file or saved for later reference.

The most important step in the laboratory procedure is to compare measurements of the actual circuit to the simulation. This will help you determine where potential errors exist in your design. For example, comparisons can help reveal inadequacies in the simulation model, or incorrect component values. After comparison with the theoretical values, you can revisit your design to improve it or prepare it for layout on a pc board.

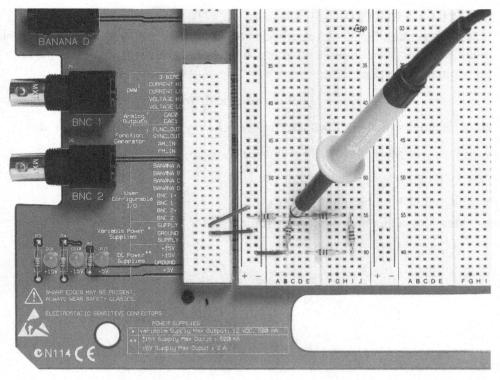

Figure I–7 The resistor network wired on an NI ELVIS

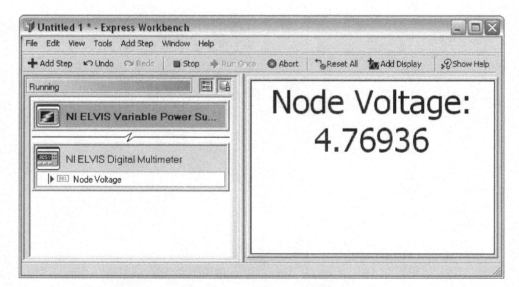

Figure I–8 NI ELVIS measurements (voltage at node A)

RESOURCES TO GET STARTED

Download the Multisim circuit files to develop in-depth understanding of circuit behavior. Simulate and analyze the circuit behavior from example programs at the end of the chapters. To download the pre-built circuit files and Multisim resources, visit:

ni.com/academic/floyd

The link provides references to the following resources to help you get started with NI Multisim:

- Download the free Multisim 30-day evaluation version
- View the Getting Started Guide to Multisim
- Learn Multisim in a 3-hour tutorial
- Discuss Multisim in an online discussion forum

Oscilloscope Guide
Analog and Digital Storage Oscilloscopes

The oscilloscope is the most widely used general-purpose measuring instrument because it allows you to see a graph of the voltage as a function of time in a circuit. Many circuits have specific timing requirements or phase relationships that can be readily measured with a two-channel oscilloscope. The voltage to be measured is converted into a visible display that is presented on a screen.

There are two basic types of oscilloscope: analog and digital. In general, they each have specific characteristics. Analog scopes are the classic "real-time" instruments that show the waveform on a cathode-ray tube (CRT). Digital oscilloscopes are rapidly replacing analog scopes because of their ability to store waveforms and because of measurement automation and many other features such as connections for computers. The storage function is so important that it is usually incorporated in the name as a Digital Storage Oscilloscope (DSO). Some higher-end DSOs can emulate an analog scope in a manner that blurs the distinction between the two types. Tektronix, for example, has a line of scopes called DPOs (Digital Phosphor Oscilloscopes) that can characterize a waveform with intensity gradients like an analog scope and gives the benefits of a digital oscilloscope for measurement automation.

Analog and digital scopes have similar functions, and the basic controls are essentially the same for both types (although certain enhanced features are not). In the descriptions that follow, the analog scope is introduced first to familiarize you with basic controls, then a specific digital storage oscilloscope is described (the Tektronix TDS1000 series).

ANALOG OSCILLOSCOPES
Block Diagram
The analog oscilloscope contains four functional blocks, as illustrated in Figure I–9. Shown within these blocks are the most important typical controls found on nearly all oscilloscopes.

Each of two input channels is connected to the vertical section, which can be set to attenuate or amplify the input signals to provide the proper voltage level to the vertical deflection plates of the CRT. In a dual-trace oscilloscope (the most common type), an electronic switch rapidly switches between channels to send one or the other to the display section.

The trigger section samples the input waveform and sends a synchronizing trigger signal at the proper time to the horizontal section. The trigger occurs at the same relative time, thus superimposing each succeeding trace on the previous trace. This action causes a repetitive signal to stop, allowing you to examine it.

The horizontal section contains the time-base (or sweep) generator, which produces a linear ramp, or "sweep," waveform that controls the rate the beam moves across the screen. The horizontal position of the beam is proportional to the time that elapsed from the start of the sweep, allowing the horizontal axis to be calibrated in units of time. The output of the horizontal section is applied to the horizontal deflection plates of the CRT.

Finally, the display section contains the CRT and beam controls. It enables the user to obtain a sharp presentation with the proper intensity. The display section usually contains other features such as a probe compensation jack and a beam finder.

Controls
Generally, controls for each section of the oscilloscope are grouped together according to function. Frequently, there are color clues to help you identify groups of controls. Details of these controls are

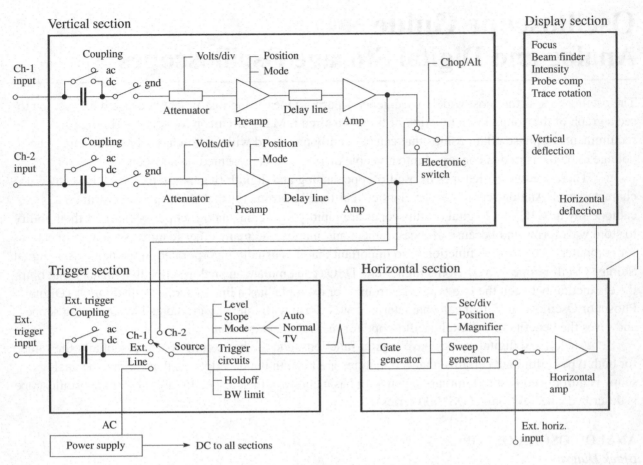

Figure I–9   Block diagram of a basic analog oscilloscope.

explained in the operator's manual for the oscilloscope; however, a brief description of frequently used controls is given in the following paragraphs. The important controls are shown on the block diagram of Figure I–9.

Display Controls   The display system contains controls for adjusting the electron beam, including FOCUS and INTENSITY controls. FOCUS and INTENSITY are adjusted for a comfortable viewing level with a sharp focus. The display section may also contain the BEAM FINDER, a control used in combination with the horizontal and vertical POSITION controls to bring the trace on the screen. Another control over the beam intensity is the z-axis input. A control voltage on the z-axis input can be used to turn the beam on or off or adjust its brightness. Some oscilloscopes also include the TRACE ROTATION control in the display section. TRACE ROTATION is used to align the sweep with a horizontal graticule line. This control is usually adjusted with a screwdriver to avoid accidental adjustment. Usually a PROBE COMP connection point is included in the display group of controls. Its purpose is to allow a quick qualitative check on the frequency response of the probe-scope system.

Vertical Controls   The vertical controls include the VOLTS/DIV (vertical sensitivity) control and its vernier, the input COUPLING switch, and the vertical POSITION control. There is a duplicate set of these controls for each channel and various switches for selecting channels or other vertical operating modes. The vertical inputs are connected through a selectable attenuator to a high input impedance dc amplifier. The VOLTS/DIV control on each channel selects a combination of attenuation/gain to determine the vertical sensitivity. For example, a low-level signal will need more gain and less attenuation than a higher level signal. The vertical sensitivity is adjusted in fixed VOLTS/DIV increments to allow the user to make calibrated voltage measurements. In addition, a concentric vernier control is usually provided to allow a continuous range of sensitivity. This knob must be in the detent (calibrated) position to make voltage measurements. The detent position can be felt by the user as the knob is turned because the knob tends to "lock" in the detent position. Some oscilloscopes have a warning light or message when the vernier is not in its detent position.

The input coupling switch is a multiple-position switch that can be set for AC, GND, or DC and sometimes includes a 50 Ω position. The GND position of the switch internally disconnects the signal from the scope and grounds the input amplifier. This position is useful if you want to set a ground reference level on the screen for measuring the dc component of a waveform. The AC and DC positions are high-impedance inputs, typically 1 MΩ shunted by 15 pF of capacitance. High-impedance inputs are useful for general probing at frequencies below about 1 MHz. At higher frequencies, the shunt capacitance can load the signal source excessively, causing measurement error. Attenuating divider probes are good for high-frequency probing because they have very high impedance (typically 10 MΩ) with very low shunt capacitance (as low as 2.5 pF).

The AC position of the coupling switch inserts a series capacitor before the input attenuator, causing dc components of the signal to be blocked. This position is useful if you want to measure a small ac signal riding on top of a large dc signal—power supply ripple, for example. The DC position is used when you want to view both the AC and DC components of a signal. This position is best when viewing digital signals because the input *RC* circuit forms a differentiating network. The AC position can distort the digital waveform because of this differentiating circuit. The 50 Ω position places an accurate 50 Ω load to ground. This position provides the proper termination for probing in 50 Ω systems and reduces the effect of a variable load which can occur in high impedance termination. The effect of source loading must be taken into account when using a 50 Ω input. It is important not to overload the 50 Ω input because the resistor is normally rated for only 2 W, implying a maximum of 10 V of signal can be applied to the input.

The vertical POSITION control varies the dc voltage on the vertical deflection plates, allowing you to position the trace anywhere on the screen. Each channel has its own vertical POSITION control, enabling you to separate the two channels on the screen. You can use vertical POSITION when the coupling switch is in the GND position to set an arbitrary level on the screen as ground reference.

There are two types of dual-channel oscilloscope: dual beam and dual trace. A dual-beam oscilloscope has two independent beams in the CRT and independent vertical deflection systems, allowing both signals to be viewed at the same time. A dual-trace oscilloscope has only one beam and one deflection system; it uses electronic switching to show the two signals. Dual-beam oscilloscopes are generally restricted to high-performance research instruments and are much more expensive than dual-trace oscilloscopes. The block diagram in Figure I–9 is for a typical dual-trace oscilloscope.

A dual-trace oscilloscope has user controls labeled CHOP or ALTERNATE to switch the beam between the channels so that the signals appear to occur simultaneously. The CHOP mode rapidly switches the beam between the two channels at a fixed high speed rate, so the two channels appear to be displayed at the same time. The ALTERNATE mode first completes the sweep for one of the channels and then displays the other channel on the next (or alternate) sweep. When viewing slow signals, the CHOP mode is best because it reduces the flicker that would otherwise be observed. High-speed signals can usually be observed best in ALTERNATE mode to avoid seeing the chop frequency.

Another feature on most dual-trace oscilloscopes is the ability to show the algebraic sum and difference of the two channels. For most measurements, you should have the vertical sensitivity

(VOLTS/DIV) on the same setting for both channels. You can use the algebraic sum if you want to compare the balance on push-pull amplifiers, for example. Each amplifier should have identical out-of-phase signals. When the signals are added, the resulting display should be a straight line, indicating balance. You can use the algebraic difference when you want to measure the waveform across an ungrounded component. The probes are connected across the ungrounded component with probe ground connected to circuit ground. Again, the vertical sensitivity (VOLTS/DIV) setting should be the same for each channel. The display will show the algebraic difference in the two signals. The algebraic difference mode also allows you to cancel any unwanted signal that is equal in amplitude and phase and is common to both channels.

Dual-trace oscilloscopes also have an X-Y mode, which causes one of the channels to be graphed on the X-axis and the other channel to be graphed on the Y-axis. This is necessary if you want to change the oscilloscope base line to represent a quantity other than time. Applications include viewing a transfer characteristic (output voltage as a function of input voltage), swept frequency measurements, or showing Lissajous figures for phase measurements. Lissajous figures are patterns formed when sinusoidal waves drive both channels and are described in Experiment 24, For Further Investigation.

Horizontal Controls   The horizontal controls include the SEC/DIV control and its vernier, the horizontal magnifier, and the horizontal POSITION control. In addition, the horizontal section may include delayed sweep controls. The SEC/DIV control sets the sweep speed, which controls how fast the electron beam is moved across the screen. The control has a number of calibrated positions divided into steps of 1-2-5 multiples which allow you to set the exact time interval that you view the input signal. For example, if the graticule has 10 horizontal divisions and the SEC/DIV control is set to 1.0 ms/div, then the screen will show a total time of 10 ms. The SEC/DIV control usually has a concentric vernier control that allows you to adjust the sweep speed continuously between the calibrated steps. This control must be in the detent position in order to make calibrated time measurements. Many scopes are also equipped with a horizontal magnifier that affects the time base. The magnifier increases the sweep time by the magnification factor, giving you increased resolution of signal details. Any portion of the original sweep can be viewed using the horizontal POSITION control in conjunction with the magnifier. This control actually speeds the sweep time by the magnification factor and therefore affects the calibration of the time base set on the SEC/DIV control. For example, if you are using a $10\times$ magnifier, the SEC/DIV dial setting must be divided by 10.

Trigger Controls   The trigger section is the source of most difficulties when learning to operate an oscilloscope. These controls determine the proper time for the sweep to begin in order to produce a stable display. The trigger controls include the MODE switch, SOURCE switch, trigger LEVEL, SLOPE, COUPLING, and variable HOLDOFF controls. In addition, the trigger section includes a connector for applying an EXTERNAL trigger to start the sweep. Trigger controls may include HIGH or LOW FREQUENCY REJECT switches and BANDWIDTH LIMITING.

The MODE switch is a multiple-position switch that selects either AUTO or NORMAL (sometimes called TRIGGERED) and may have other positions such as TV or SINGLE sweep. In the AUTO position, the trigger generator selects an internal oscillator that will trigger the sweep generator as long as no other trigger is available. This mode ensures that a sweep will occur even in the absence of a signal because the trigger circuits will "free-run" in this mode. This allows you to obtain a baseline for adjusting ground reference level or for adjusting the display controls. In the NORMAL or TRIGGERED mode, a trigger is generated from one of three sources selected by the SOURCE switch—the INTERNAL signal, an EXTERNAL trigger source, or the AC LINE. If you are using the internal signal to obtain a trigger, the normal mode will provide a trigger only if a signal is present and other trigger conditions (level, slope) are met. This mode is more versatile than AUTO as it can provide stable triggering for very low to very high frequency signals. The TV position is used for synchronizing either television fields or lines and SINGLE is used primarily for photographing the display.

The trigger LEVEL and SLOPE controls are used to select a specific point on either the rising or falling edge of the input signal for generating a trigger. The trigger SLOPE control determines which edge will generate a trigger, whereas the LEVEL control allows the user to determine the voltage level on the input signal which will start the sweep circuits.

The SOURCE switch selects the trigger source—either from the CH-1 signal, the CH-2 signal, an EXTERNAL trigger source, or the AC LINE. In the CH-1 position, a sample of the signal from channel-1 is used to start the sweep. In the EXTERNAL position, a time-related external signal is used for triggering. The external trigger can be coupled with either AC or DC COUPLING. The trigger signal can be coupled with AC COUPLING if the trigger signal is riding on a dc voltage. DC COUPLING is used if the triggers can occur at a frequency of less than about 20 Hz. The LINE position causes the trigger to be derived from the ac power source. This synchronizes the sweep with signals that are related to the power line frequency.

The variable HOLDOFF control allows you to exclude otherwise valid triggers until the holdoff time has elapsed. For some signals, particularly complex waveforms or digital pulse trains, obtaining a stable trigger can be a problem. This can occur when one or more valid trigger points occur before the signal repetition time. If every event that the trigger circuits qualified as a trigger were allowed to start a sweep, the display could appear to be unsynchronized. By adjusting the variable HOLDOFF control, the trigger point can be made to coincide with the signal-repetition point.

OSCILLOSCOPE PROBES
Signals should always be coupled into the oscilloscope through a probe. A probe is used to pick off a signal and couple it to the input with a minimum loading effect on the circuit under test. Various types of probes are provided by manufacturers but the most common type is a 10:1 attenuating probe that is shipped with most general-purpose oscilloscopes. These probes have a short ground lead that should be connected to a nearby circuit ground point to avoid oscillation and power line interference. The ground lead makes a mechanical connection to the test circuit and passes the signal through a flexible, shielded cable to the oscilloscope. The shielding helps protect the signal from external noise pickup.

Begin any session with the oscilloscope by checking the probe compensation on each channel. Adjust the probe for a flat-topped square wave while observing the scope's calibrator output. This is a good signal to check the focus and intensity and verify trace alignment. Check the front-panel controls for the type of measurement you are going to make. Normally, the variable controls (VOLTS/DIV and SEC/DIV) should be in the calibrated (detent) position. The vertical coupling switch is usually placed in the DC position unless the waveform in which you are interested has a large dc offset. Trigger holdoff should be in the minimum position unless it is necessary to delay the trigger to obtain a stable sweep.

DIGITAL STORAGE OSCILLOSCOPES
Block Diagram
The digital storage oscilloscope (DSO) uses a fast analog-to-digital converter (ADC) on each channel (typically two or four channels) to convert the input voltage into numbers that can be stored in a memory. The digitizer samples the input at a uniform rate called the sample rate; the optimum sample rate depends on the speed of the signal. The process of digitizing the waveform has many advantages for accuracy, triggering, viewing hard-to-see events, and for waveform analysis. Although the method of acquiring and displaying the waveform is quite different than analog scopes, the basic controls on the instrument are similar.

A block diagram of the basic DSO is shown in Figure I–10. As you can see, functionally, the block diagram is similar to the analog scope. As in the analog oscilloscope, the vertical and horizontal controls include position and sensitivity, which are used to set up the display for the proper scaling.

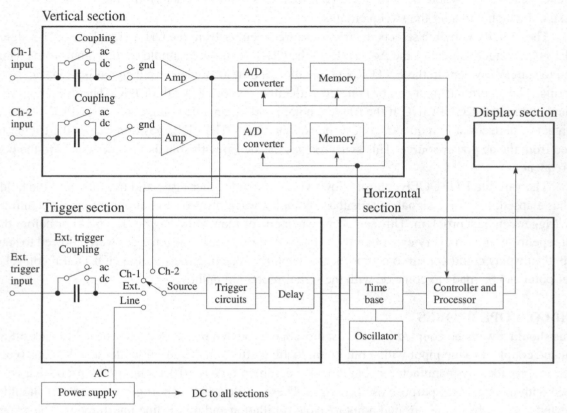

Figure I–10 Block diagram of a basic digital storage oscilloscope.

Specifications   Important parameters with DSOs include the resolution, maximum digitizing rate, and the size of the acquisition memory as well as the available analysis options. The resolution is determined by the number of bits digitized by the ADC. A low-resolution DSO may use only six bits (one part in 64). A typical DSO may use 8 bits, with each channel sampled simultaneously. High-end DSOs may use 12 bits. The maximum digitizing rate is important to capture rapidly changing signals; typically the maximum rate is 1 Gsample/s. The size of the memory determines the length of time the sample can be taken; it is also important in certain waveform measurement functions.

Triggering   One useful feature of digital storage oscilloscopes is their ability to capture waveforms either before or after the trigger event. Any segment of the waveform, either before or after the trigger event, can be captured for analysis. **Pretrigger capture** refers to acquisition of data that occurs *before* a trigger event. This is possible because the data are digitized continuously, and a trigger event can be selected to stop the data collection at some point in the sample window. With pretrigger capture, the scope can be triggered on the fault condition, and the signals that preceded the fault condition can be observed. For example, troubleshooting an occasional glitch in a system is one of the most difficult troubleshooting jobs; by employing pretrigger capture, trouble leading to the fault can be analyzed. A similar application of pretrigger capture is in material failure studies where the events leading to failure are most interesting but the failure itself causes the scope triggering.

Besides pretrigger capture, posttriggering can also be set to capture data that occur some time after a trigger event. The record that is acquired can begin after the trigger event by some amount of time or by a specific number of events as determined by a counter. A low-level response to a strong stimulus signal is an example of when posttriggering is useful.

A Specific DSO   Because of the large number of functions that can be accomplished by even basic DSOs, manufacturers have largely replaced the plethora of controls with menu options, similar to computer menus and detailed displays that show the controls as well as measurement parameters. CRTs have been replaced by liquid crystal displays, similar to those on laptop computers. As an example, the display for a Tektronix TDS1000 and 2000 series digital storage oscilloscope is shown in Figure I–11. Although this is a basic scope, the information available to the user right on the display is impressive.

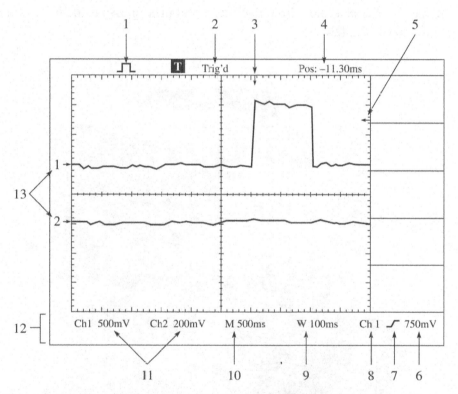

Figure I–11   The display area for Tektronix TDS1000 and 2000 series oscilloscope

The numbers on the display in Figure I–11 refer to the following parameters:
1.   Icon display shows acquisition mode.
     Sample mode
     Peak detect mode
     Average mode
2.   Trigger status shows if there is an adequate trigger source or if the acquisition is stopped.
3.   Marker shows horizontal trigger position. This also indicates the horizontal position since the Horizontal Position control actually moves the trigger position horizontally.

13

4. Trigger position display shows the difference (in time) between the center graticule and the trigger position. Center screen equals zero.

5. Marker shows trigger level.

6. Readout shows numeric value of the trigger level.

7. Icon shows selected trigger slope for edge triggering.

8. Readout shows trigger source used for triggering.

9. Readout shows window zone time-base setting.

10. Readout shows main time-base setting.

11. Readout shows channels 1 and 2 vertical scale factors.

12. Display area shows on-line messages momentarily.

13. On-screen markers show the ground reference points of the displayed waveforms. No marker indicates the channel is not displayed.

A front view of the TDS1000 and 2000 series is shown in Figure I–12. Operation is similar to that of an analog scope except more of the functions are menu controlled; in the TDS1000 and 2000 series, 12 different menus are accessed to select various controls and options. For example, the MEASURE function brings up a menu that the user can select from five automated measurements including voltage, frequency, period, and averaging to name a few.

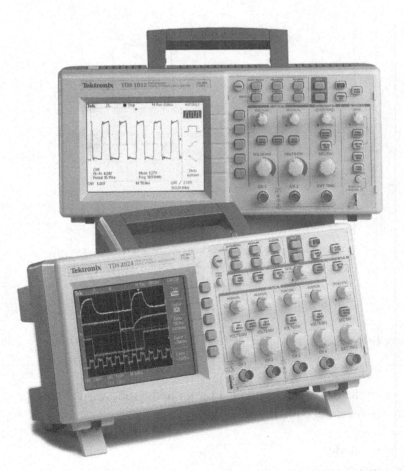

Figure I–12 The Tektronix TDS1000 and 2000 series oscilloscope (courtesy of Tektronix, Inc.).

The Technical Report

EFFECTIVE WRITING

The purpose of technical reports is to communicate technical information in a way that is easy for the reader to understand. Effective writing requires that you know your reader's background. You must be able to put yourself in the reader's place and anticipate what information you must convey to have the reader understand what you are trying to say. When you are writing experimental results for a person working in your field, such as an engineer, your writing style may contain words or ideas that are unfamiliar to a layperson. If your report is intended for persons outside your field, you will need to provide background information.

WORDS AND SENTENCES

You need to choose words that have clear meaning to a general audience or define every term, including acronyms, that does not have a well-established meaning. Keep sentences short and to the point. Short sentences are easier for the reader to comprehend. Avoid stringing a series of adjectives or modifiers together. For example, the meaning of this figure caption is unclear:

> *Operational amplifier constant-current source schematic*

The noun *schematic* is described by two modifiers, each of which has its own modifier. By changing the order and adding natural connectors such as *of, using,* and *an,* the meaning can be clarified:

> *Schematic of a constant-current source using an operational amplifier*

PARAGRAPHS

Paragraphs need to contain a unit of thought. Excessively long paragraphs suffer from the same weakness that afflict overly long sentences. The reader is asked to digest too much material at once, causing comprehension to diminish. Paragraphs should organize your thoughts in a logical format. Look for natural breaks in your ideas. Each paragraph should have one central idea and contribute to the development of the entire report.

Good organization is the key to a well-written report. Outlining in advance will help organize your ideas. The use of headings and subheadings for paragraphs or sections can help steer the reader through the report. Subheadings also prepare the reader for what is ahead and make the report easier to understand.

FIGURES AND TABLES

Figures and tables are effective ways to present information. Figures should be kept simple and to the point. Often a graph can make clear the relationship of data. Comparisons of different data drawn on the same graph make the results more obvious to the reader. Figures should be labeled with a figure number and a brief label. Don't forget to label both axes of graphs.

Data tables are useful for presenting data. Usually data presented in a graph or figure should not also be included in a data table. Data tables should be labeled with a table number and short title. The data table should contain enough information that its meaning is clear to the reader without having to refer to the text. If the purpose of the table is to compare information, then form the data in columns rather than rows. Information in columns is easier for people to compare. Table footnotes are a useful method of clarifying some point about the data. Footnotes should appear at the bottom of the table with a key to where the footnote applies.

Data should appear throughout your report in consistent units of measurement. Most sciences use the metric system; however, the English (or customary) system is still sometimes used. The metric system uses derived units that are cgs (centimeter-gram-second) or mks (meter-kilogram-second). It is best to use consistent metric units throughout your report.

Tabular data should be shown with a number of significant digits consistent with the precision of the measurement. Significant digits are discussed in the *Summary of Theory* for Experiment 1.

Reporting numbers using powers of 10 can be a sticky point with reference to tables. Table I–1 shows four methods of abbreviating numbers in tabular form. The first column is unambiguous; the number is presented in conventional form. This requires more space than if the information is presented in scientific notation. In column 2, the same data are shown with a metric prefix used for the unit. In column 3, the power of 10 is shown. Each of the first three columns shows the measurement unit and is not subject to misinterpretation. Column 4, on the other hand, is wrong. In this case, the author is trying to tell us what operation was performed on the numbers to obtain the values in the column. This is incorrect because the column heading should contain the unit of measurement for the numbers in the column.

Table I–1 Reporting numbers in tabular data.

Column 1	Column 2	Column 3	Column 4
Resistance ohms	Resistance $k\Omega$	Resistance $\times 10^3$ ohms	Resistance ohms $\times 10^{-3}$
470,000	470	470	470
8,200	8.2	8.2	8.2
1,200,000	1,200	1,200	1,200
330	0.33	0.33	0.33
	— Correct —		Wrong

SUGGESTED FORMAT

1. *Title.* A good title needs to convey the substance of your report by using key words that provide the reader with enough information to decide if the report should be investigated further.
2. *Contents.* Key headings throughout the report are listed with page numbers.
3. *Abstract.* The abstract is a brief summary of the work with principal facts and results stated in concentrated form. It is a key factor in helping a reader to determine if he or she should read further.
4. *Introduction.* The introduction orients a reader. It should briefly state what you did and give the reader a sense of the purpose of the report. It may tell the reader what to expect and briefly describe the report's organization.
5. *Body of the report.* The report can be made clearer to the reader if you use headings and subheadings to mark major divisions through your report. The headings and subheadings can be generated from the outline of your report. Figures and tables should be labeled and referenced in the body of the report.

6.   *Conclusion.* The conclusion summarizes important points or results. It may refer to figures or tables previously discussed in the body of the report to add emphasis to significant points. In some cases, the primary reasons for the report are contained within the body and a conclusion is deemed to be unnecessary.

7.   *References.* References are cited to enable the reader to find information used in developing your report or work that supports your report. The references should include names of all authors, in the order shown in the original document. Use quotation marks around portions of a complete document such as a journal article or a chapter of a book. Books, journals, or other complete documents should be underlined. Finally, list the publisher, city, date, and page numbers.

1 Metric Prefixes, Scientific Notation, and Graphing

Name _____
Date _____
Class _____

READING
Text, Sections 1–1 through 1–5

OBJECTIVES
After performing this experiment, you will be able to:
1. Convert standard form numbers to scientific and engineering notation.
2. Measure quantities using a metric prefix.
3. Prepare a linear graph and plot a family of curves on the graph.

MATERIALS NEEDED
Scientific calculator
Metric ruler

SUMMARY OF THEORY
The basic electrical quantities encompass a very large range of numbers—from the very large to the very small. For example, the frequency of an FM radio station can be over 100 million hertz (Hz), and a capacitor can have a value of 10 billionths of a farad (F). To express very large and very small numbers, scientific (powers of ten) notation and metric prefixes are used. Metric prefixes are based on the decimal system and stand for powers of ten. They are widely used to indicate a multiple or submultiple of a measuring unit.

    Scientific notation is a means of writing any quantity as a number between 1 and 10 times a power of 10. The power of 10 is called the exponent. It simply shows how many places the decimal point must be shifted to express the number in its standard form. If the exponent is positive, the decimal point must be shifted to the right to write the number in standard form. If the exponent is negative, the decimal point must be shifted to the left. Note that $10^0 = 1$, so an exponent of zero does not change the original number.

    Exponents that are a multiple of 3 are much more widely used in electronics work than exponents which are not multiples of 3. Numbers expressed with an exponent that is a multiple of 3 are said to be expressed in **engineering notation.** Engineering notation is particularly useful in electronics work because of its relationship to the most widely used metric prefixes. Some examples of numbers written in standard form, scientific notation, and engineering notation are shown in Table 1–1.

Table 1–1

Standard Form	Scientific Notation	Engineering Notation
12,300	1.23×10^4	12.3×10^3
506	5.06×10^2	0.506×10^3
8.81	8.81×10^0	8.81×10^0
0.0326	3.26×10^{-2}	32.6×10^{-3}
0.000 155	1.55×10^{-4}	155×10^{-6}

Numbers expressed in engineering notation can be simplified by using metric prefixes to indicate the appropriate power of ten. In addition, prefixes can simplify calculations. You can perform arithmetic operations on the significant figures of a problem and determine the answer's prefix from those used in the problem. For example, 4.7 kΩ + 1.5 kΩ = 6.2 kΩ. The common metric prefixes used in electronics and their abbreviations are shown in Table 1–2. The metric prefixes representing engineering notation are shown. Any number can be converted from one prefix to another (or no prefix) using the table. Write the number to be converted on the line with the decimal under the metric prefix that appears with the number. The decimal point is then moved directly under any other line, and the metric prefix immediately above the line is used. The number can also be read in engineering notation by using the power of ten shown immediately above the line.

Table 1–2

Power of 10:	10^9	10^6	10^3	10^0	10^{-3}	10^{-6}	10^{-9}	10^{-12}
Metric symbol:	G	M	k		m	μ	n	p
Metric prefix:	giga	mega	kilo		milli	micro	nano	pico

0 ' 000 ' 000 ' 000 . 000 ' 000 ' 000 ' 000 ' 0

Example 1:

Convert 12,300,000 Ω to a number with an M prefix:

Metric prefix:	giga	mega	kilo		milli	micro	nano	pico

0 ' 000 ' 000 ' 000 . 000 ' 000 ' 000 ' 000 ' 0
    12   300   000 . Ω

= 1 2 . 3 MΩ

Example 2:

Change 10,000 pF to a number with a μ prefix:

Metric prefix:	giga	mega	kilo		milli	micro	nano	pico

0 ' 000 ' 000 ' 000 . 000 ' 000 ' 000 ' 000 ' 0
                                    10 000 . pF

=. 0 1 0 μF

CALCULATORS

Scientific calculators have the ability to process numbers that are written in exponential form. In addition, scientific calculators can perform trig functions, logarithms, roots, and other math functions. To enter numbers in scientific notation on most calculators, the base number (called the *mantissa*) is first entered. If the number is negative, the +/− key is pressed. Next the exponent is entered by pressing the EE (or

EXP) key, followed by the power of ten.[1] If the exponent is negative, the $+/-$ key is pressed. Arithmetic can be done on the calculator with numbers in scientific notation mixed with numbers in standard form. On many calculators, such as the TI-86, there is an engineering mode. (The TI-86 is placed in engineering mode by pressing $\boxed{2^{nd}}$ $\boxed{\text{MODE}}$ and selecting $\boxed{\text{Eng}}$). Engineering mode is particularly useful for electronics calculations because of the direct correlation to the metric prefixes in Table 1–2.

SIGNIFICANT DIGITS

When a measurement contains approximate data, those digits known to be correct are called *significant digits*. Zeros that are used only for locating the decimal place are *not* significant, but those that are part of the measured quantity are significant. When reporting a measured value, the least significant uncertain digit may be retained, but all other uncertain digits should be discarded. It is *not* correct to show either too many or too few digits. For example, it is not valid to retain more than three digits when using a meter that has three digit resolution, nor is it proper to discard valid digits, even if they are zeros. For example, if you set a power supply to the nearest hundredth of a volt, then the recorded voltage should be reported to the hundredth place (3.00 V is correct, but 3 V is incorrect). For laboratory work in this course, you should normally be able to measure and retain three significant digits.

To find the number of significant digits in a given number, ignore the decimal point and count the number of digits from left to right, starting with the first nonzero digit and ending with the last digit to the right. All digits counted are significant except zeros at the right end of the number. A zero on the right end of a number is significant *only* if it is to the right of the decimal point; otherwise it is uncertain. For example, 43.00 contains four significant digits. The whole number 4300 may contain two, three, or four significant digits. In the absence of other information, the significance of the right-hand zeros is uncertain, and these digits cannot be assumed to be significant. To avoid confusion, numbers such as these should be reported using scientific notation. For example, the number 2.60×10^3 contains three significant figures and the number 2.600×10^3 contains four significant figures.

Rules for determining if a reported digit is significant are as follows.
1. Nonzero digits are always considered to be significant.
2. Zeros to the left of the first nonzero digit are never significant.
3. Zeros between nonzero digits are always significant.
4. Zeros at the right of a nonzero digit in a decimal number are always significant.
5. Zeros at the right end of a whole number are uncertain. Whole numbers should be reported in scientific or engineering notation to clarify the significant figures.

GRAPHS

A graph is a visual tool that can quickly convey to the reader the relationship between variables. The eye can discern trends in magnitude or slope more easily on a graph than from tabular data. Graphs are widely used in experimental work to present information because they enable the reader to discern variations in magnitude, slope, and direction between two quantities. In this manual, you will graph data in many experiments.

---

[1]Note that when you are entering numbers in scientific notation on the calculator, it is not necessary to enter the base ten, only the exponent.

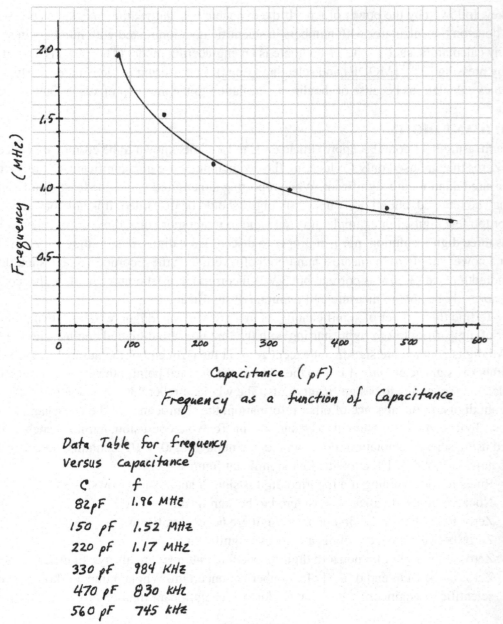

Capacitance (pF)

Frequency as a function of Capacitance

Data Table for frequency
Versus capacitance

C	f
82 pF	1.96 MHz
150 pF	1.52 MHz
220 pF	1.17 MHz
330 pF	984 KHz
470 pF	830 KHz
560 pF	745 KHz

Figure 1–1

The six steps in preparing a graph were described in the Introduction to the Student on page x. Figure 1–1 illustrates a set of data and a linear graph of the data. Notice that the six steps are followed in preparing this graph. For this particular example, the steps are:

Step 1: Select a linear scale. The independent variable (capacitance) is on the *x*-axis. Choose 20 pF/div on the *x*-axis and 0.1 MHz/div on the *y*-axis to fit all of the data on the plot.

Step 2: Number the major divisions. For the *x*-axis, select 100 pF increments; for the *y*-axis, select 0.5 MHz.

Step 3: Add labels and units to each axis. Place capacitance (pF) and Frequency (MHz) along their respective axis.

Step 4: Transfer data points from the Data table to the plot.

Step 5: Draw a smooth curve. Notice that it is not necessary to touch every data point.

Step 6: Add a title. In this case, the title describes the plot "Frequency as a function of capacitance."

PROCEDURE

1. Many of the dials and controls of laboratory instruments are labeled with metric prefixes. Check the controls on instruments at your lab station for metric prefixes. For example, check the SEC/DIV control on your oscilloscope. This control usually has more than one metric prefix associated with the switch positions. Meters are also frequently marked with metric prefixes. Look for others and list the instrument, control, metric unit and its meaning in Table 1–3. There are many possible answers. The first line of Table 1–3 has been completed as an example.

Table 1–3

Instrument	Control	Metric Unit	Meaning
Oscilloscope	SEC/DIV	ms	10^{-3} s

Table 1–4

Dimension	Length in Millimeters	Length in Meters
A	7.2 mm	7.2×10^{-3} m
B		
C		
D		
E		
F		
G		

2. The actual sizes of several electronic components are shown in Figure 1–2. Measure the quantities shown with a bold letter using a metric ruler. Report in Table 1–4 the length in millimeters of each lettered quantity. Then rewrite the measured length as the equivalent length in meters and record your results in Table 1–4. The first line of the table has been completed as an example. (Lengths are approximate.)

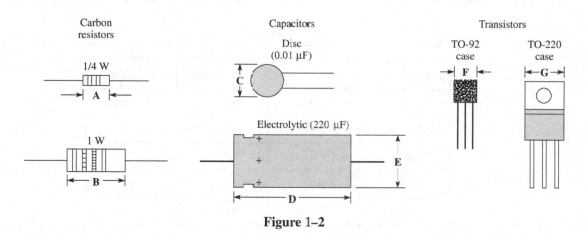

Figure 1–2

Table 1–5

Number	Scientific Notation	Engineering Notation	Metric Value
0.0829 V	8.29×10^{-2} V	82.9×10^{-3} V	82.9 mV
48,000 Hz			
2,200,000 Ω			
0.000 015 A			
7,500 W			
0.000 000 033 F			
270,000 Ω			
0.000 010 H			

Table 1–6

Metric Value	Engineering Notation
100 pF	100×10^{-12} F
12 kV	
85.0 μA	
50 GHz	
33 kΩ	
250 mV	
7.8 ns	
2.0 MΩ	

3.   Rewrite the numbers in Table 1–5 in scientific notation, engineering notation, and using one of the engineering metric prefixes. The first line has been completed as an example.

4.   Convert the metric values listed in Table 1–6 to engineering notation. The first line has been completed as an example.

5.   Metric prefixes are useful for solving problems without having to key in the exponent on your calculator. For example, when a milli prefix (10^{-3}) is multiplied by a kilo prefix (10^{+3}), the metric prefixes cancel and the result has only the measuring unit. As you become proficient with these prefixes, math operations are simplified and fewer keystrokes are required in solving the problem with a calculator. To practice this, determine the metric prefix for the answer when each operation indicated in Table 1–7 is performed. The first line is shown as an example.

Table 1–7

Metric Unit in Operand	Mathematical Operation	Metric Unit in Operand		Metric Unit in Result
milli	multiplied by	milli	=	micro
kilo	multiplied by	micro	=	
nano	multiplied by	kilo	=	
milli	multiplied by	mega	=	
micro	divided by	nano	=	
micro	divided by	pico	=	
pico	divided by	pico	=	
milli	divided by	mega	=	

6.   This step is to provide you with practice in graphing and in presenting data. Table 1–8 lists inductance data for 16 different coils wound on identical iron cores. There are three variables in this problem: the length of the coil (*l*) given in centimeters (cm), the number of turns, *N*, and the inductance, *L*, given in millihenries (mH). Since there are three variables, we will hold one constant and plot the data using the remaining two variables. This procedure shows how one variable relates to the other. Start by plotting the length (first column) as a function of inductance (last column) for coils that have 400 turns. Use Plot 1–1. The steps in preparing a graph are given in the *Introduction to the Student* and reviewed with an example in the Summary of Theory, page 22.

Table 1–8   Inductance, *L*, of coils wound on identical iron cores (mH).

Length, *l* (cm)	Number of Turns, *N* (t)			
	100	200	300	400
2.5	3.9	16.1	35.8	64.0
5.5	1.7	7.5	16.1	29.3
8.0	1.2	5.1	11.4	19.8
12.0	0.8	3.3	7.5	13.1

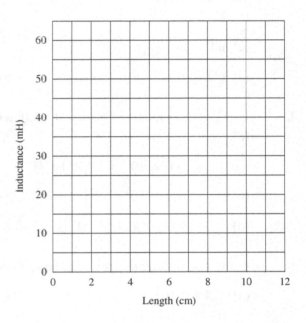

Plot 1–1

7.   On the same plot, graph the data for the 300 turn coils, then the 200 turn and 100 turn coils. Use a different symbol for each set of data. The resulting graph is a family of curves that give a quick visual indication of the relationship among the three variables.

CONCLUSION
Example
This experiment included exercises in measuring and plotting. The data in steps 6 and 7 show that, for a given number of turns, the inductance is smaller for a longer coil. The four lines indicate that inductance increases as the number of turns is larger.

EVALUATION AND REVIEW QUESTIONS

1. For each metric prefix and unit shown, write the abbreviation of the metric prefix with the unit symbol:
   (a) kilowatt   (b) milliampere

   (c) picofarad   (d) nanosecond

   (e) megohm   (f) microhenry

2. Write the metric prefix and unit name for each of the abbreviations shown:
   (a) MW   (b) nA

   (c) μJ   (d) mV

   (e) kΩ (f) GHz

3. Using your calculator, perform the following operations:
   (a) $(3.6 \times 10^4)(8.8 \times 10^{-4})$

   (b) $(-4.0 \times 10^{-6})(2.7 \times 10^{-1})$

   (c) $(-7.5 \times 10^2)(-2.5 \times 10^{-5})$

   (d) $(56 \times 10^3)(9.0 \times 10^{-7})$

4. Using your calculator, perform the following operations:

   (a) $\dfrac{(4.4 \times 10^9)}{(-7.0 \times 10^3)}$ (b) $\dfrac{(3.1 \times 10^2)}{(41 \times 10^{-6})}$

   (c) $\dfrac{(-2.0 \times 10^4)}{(-6.5 \times 10^{-6})}$ (d) $\dfrac{(0.0033 \times 10^{-3})}{(-15 \times 10^{-2})}$

5. For each result in Question 4, write the answer as one with a metric prefix:
   (a)   (b)

   (c)   (d)

6. Summarize the six steps in preparing a linear graph:

FOR FURTHER INVESTIGATION

In steps 6 and 7, it is apparent that the data for the 100 turn coils is close to the *x*-coordinate, making it difficult to read on the same graph as the 400 turn data. A solution to this problem is to plot the data on a log-log plot. A logarithmic scale increases the resolution when data encompasses a large range of values. To help you get started, the axes have already been labeled and values assigned. Plot the data from Table 1–8 onto Plot 1–2. You should observe that each data set will plot a straight line. This result indicates the form of the equation that relates the variables is a power function.

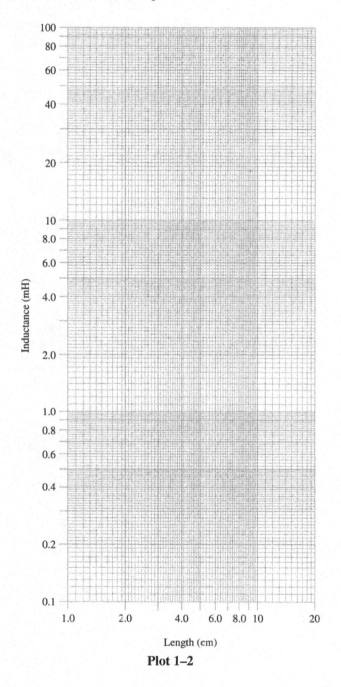

Plot 1–2

Checkup 1

REFERENCE
Text, Chapter 1; Lab manual, Experiment 1

1. The metric prefix *micro* means
   (a) 10^3 (b) 10^{-3} (c) 10^{-6} (d) 10^{-9}

2. The metric prefix *mega* means
   (a) 10^3 (b) 10^6 (c) 10^{-9} (d) 10^{12}

3. The metric prefix *nano* means
   (a) 10^3 (b) 10^{-3} (c) 10^{-6} (d) 10^{-9}

4. The number of pF in 0.01 μF is
   (a) 10   (b) 10^2 (c) 10^3 (d) 10^4

5. The number of mA in 10 A is
   (a) 10   (b) 10^2 (c) 10^3 (d) 10^4

6. One second is equal to
   (a) 10^6 ns   (b) 10^{-6} ns   (c) 10^9 ns   (d) 10^{-9} ns

7. One-fourth watt is the same as
   (a) 250 mW   (b) 0.025 mW   (c) 250 μW   (d) C.025 μW

8. The number of significant figures in the number 0.07020 is
   (a) 3   (b) 4   (c) 5   (d) 6

9. The number of significant figures in the number 2.0600×10^3 is
   (a) 2   (b) 3   (c) 4   (d) 5

10. In graphing data from an experiment, it is important to
    (a) label the axes including units   (b) connect all data points with a line
    (c) number all minor divisions   (d) include your name on the graph

11. Explain the difference between scientific notation and engineering notation.

12. What is the difference between the abscissa and the ordinate in a graph?

13. List the unit of measurement for each:
    (a) resistance   (b) capacitance
    (c) frequency   (d) inductance
    (e) voltage   (f) energy

14. Show the symbol for each of the following measurement units:
    (a) ohm   (b) farad
    (c) watt   (d) hertz
    (e) coulomb   (f) ampere

15. Express the following numbers in scientific notation as a number between 1 and 10 times 10 to
    the appropriate power:
    (a) 1050   (b) 0.0575
    (c) 251×10^2 (d) 89.0×10^{-5}
    (e) 0.000 004 91   (f) 0.0135×10^{-2}

16. Express the following numbers in engineering notation:
    (a) 0.00520   (b) 59 200
    (c) 760×10^5 (d) 19.0×10^{-4}
    (e) 1.22×10^2 (f) 0.0509×10^{-10}

17. Change each quantity from scientific notation to a number with a metric prefix:
    (a) 1.24×10^{-6} A   (b) $7.5 \times 10^3 \ \Omega$
    (c) 4.7×10^4 Hz   (d) 3.3×10^{-8} F
    (e) 2.2×10^{-12} s   (f) 9.5×10^{-2} H

18. Change each quantity as indicated:
    (a) 70 μA to amps   (b) 50 MHz to hertz
    (c) 0.010 μF to farads   (d) 5.0 W to milliwatts
    (e) 22 mV to volts   (f) 3300 pF to microfarads

19. Perform the following additions. Express answers with three significant digits:
    (a) $5.25 \times 10^3 + 4.97 \times 10^3$ (b) $9.02 \times 10^4 + 1.66 \times 10^3$
    (c) $1.00 \times 10^{-2} + 2.25 \times 10^{-2}$ (d) $4.15 \times 10^{-6} + 6.8 \times 10^{-7}$
    (e) $9.60 \times 10^{-5} + 1.95 \times 10^{-4}$ (f) $8.79 \times 10^6 + 4.85 \times 10^7$

20. What type of variable is plotted on the *x*-axis of a graph?

21. If you are plotting data on a linear graph, what is true about each division assigned to an axis?

2 Laboratory Meters and Power Supply

Name _____
Date _____
Class _____

READING

Text, Sections 2–1 through 2–7
Operator's Manual for Laboratory Multimeter and Power Supply
Circuit Simulation and Prototyping (page 1 of this manual)

OBJECTIVES

After performing this experiment, you will be able to:

1. Read analog meter scales including multiple and complex scales.
2. Operate the power supply at your lab station.
3. Explain the functions of the controls for the multimeter at your lab station. Use it to make a voltage reading.

MATERIALS NEEDED

None
For Further Investigation:
  Meter calibrator

SUMMARY OF THEORY

Work in electronic laboratories requires you to be familiar with basic instruments that you will use throughout your study of electronics. In this experiment, you will set up and measure voltage. You should review safety procedures for laboratory work before attempting to use the power supply and meters that are introduced in this experiment. You will only use low voltages, but you should still be aware of hazards when using any electrical equipment. You should never touch a "live" circuit, even if it is low voltage.

A detailed definition of voltage is given in the text in Chapter 2; a short discussion follows to get you started in this experiment with the power supply and using meters.

You can think of voltage as the "driving force" for a circuit. Technically speaking, it is not a force, but it does cause current. Electrical circuits contain a source of voltage and components connected in a manner to provide a path for current. Many components require a very stable source of constant voltage. This voltage is called dc (for direct current) and is usually supplied by a regulated dc power supply. Regulated power supplies are circuits that convert the alternating current (ac) line power into a constant output voltage in spite of changes to the input ac, the load current, or temperature.

The amount of voltage required varies widely for different circuits, depending on the type of circuit and the power levels. It is important that the power supply be set up *prior* to connecting it to the circuit to avoid damaging sensitive components. Most of the time the power supply voltage is checked with a meter called a *voltmeter*. The measurement of voltage is also important for determining circuit performance in many cases. Another important electrical parameter is resistance, which is the opposition to current, measured in ohms. The meter that measures resistance is called the *ohmmeter*. Voltage and resistance measurements will be the main focus of this experiment because they are often measured with the same meter. Other measurements will be described in later experiments.

Many electrical quantities are measured with meters. The schematic symbol for a basic meter is shown in Figure 2–1. The meter function is shown on the schematic with a letter or symbol. One popular

31

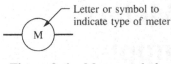

Letter or symbol to
indicate type of meter

Figure 2–1   Meter symbol.

type of meter is the *multimeter,* an instrument which combines three basic meters into one. The multimeter can measure resistance, voltage, or current and sometimes includes other types of measurements. Multimeters may be either analog or digital. An analog multimeter (VOM or <u>V</u>olt-<u>O</u>hm-<u>M</u>illiammeter) uses a pointer to indicate on a numbered scale the value of the measured quantity. A digital multimeter (or DMM) shows the measured quantity as a number. Digital multimeters are more widely used than analog multimeters because of superior performance and ease of use. Examples of a portable VOM and DMM are shown in Figure 2–2.

For voltage measurements, the multimeter needs to be set up for either ac or dc. Depending on the meter, you may need to select a *range* for the reading. Electrical quantities extend from the very small to the very large. Resistance, for example, can vary from less than 1 Ω to over 1,000,000 Ω. Meters must have some means of accommodating this large range of numbers. The position of the decimal point is determined by the range switch on the meter. The user selects an appropriate range to display the measured number. Some meters have autoranging, which means they can change ranges automatically. An autoranging meter may also have an AUTO/HOLD switch that allows the meter to either operate in the autoranging mode or to hold the last range setting.

When you use an autoranging multimeter, the multimeter will normally be in the AUTO mode. The function to be measured is selected, and the multimeter is connected to the circuit under test. The user must be careful to connect the meter correctly for the measurement to be made. Examples of how to connect an autoranging DMM for measurement of voltage and resistance are shown in Figure 2–3. There are often limitations to the ability of a meter to measure accurately, so you should be familiar with these limitations before using it.

Current measurements require special care to avoid damage to the multimeter. An ammeter must never be connected across a voltage source. When a multimeter is used to measure current, the function-

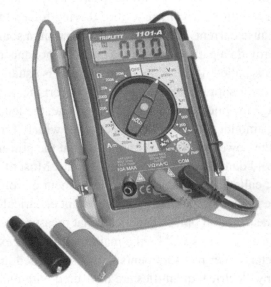

(a) VOM   (b) DMM (courtesy of Triplett Co.)

Figure 2–2

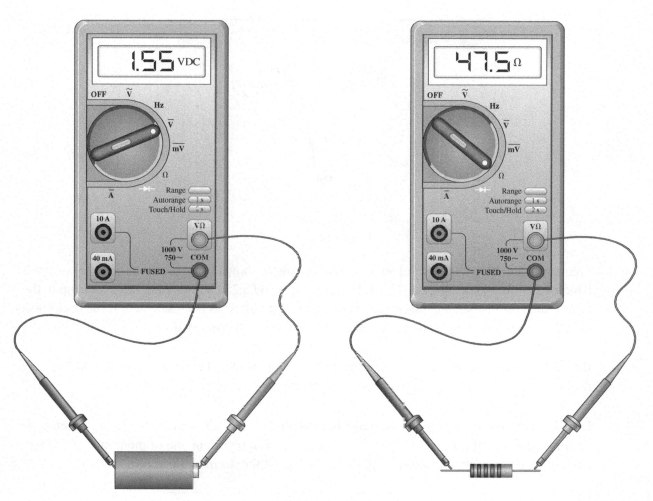

Figure 2–3

select switch is moved to the current position; the probe must be moved to a current-measuring socket before connecting the meter in the circuit. In this experiment, current will not be measured.

Many electronic measurements are made with analog meters. Analog meters can be calibrated to read almost any physical quantity, including voltage, current, power, or even nonelectrical quantities such as weight, speed, or light. The scales on analog meters may be either linear or nonlinear. They may have several scales on the same meter face. Various types of meters will be described in the Procedure section of this experiment.

PROCEDURE

1. A linear meter scale is marked in equally spaced divisions across the face of the meter. Figure 2–4 shows a linear meter scale. The major divisions, called *primary* divisions, are usually numbered. Between the primary divisions are smaller divisions called *secondary* divisions. To read this scale, note the number of secondary divisions between the numbered primary divisions and determine the value of each secondary division. The scale shown has 10 subdivisions.

   What is the value of each secondary division in Figure 2–4? _____ What is the meter reading? _____

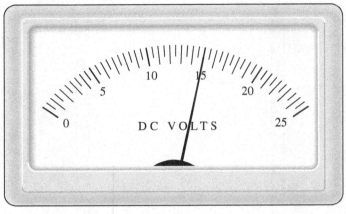

Figure 2–4

2. Frequently a meter is used for several ranges. The meter shown in Figure 2–4 could, for example, have a 2.5 V full-scale range, a 25 V full-scale range, and a 250 V full-scale range. It is up to the user to then set the decimal place, depending on which range has been selected. If the user selects the 250 V full-scale range, then there are 50 V between each primary division.

If the meter shown in Figure 2–4 is on the 250 V range, what is the value of each secondary division? _____ What is the meter reading? _____

3. Usually, meters with more than one range have several scales called *multiple* scales. A meter with multiple scales is illustrated in Figure 2–5. Each scale can represent one or more ranges. In this case, the user must choose the appropriate scale *and* set the decimal place.

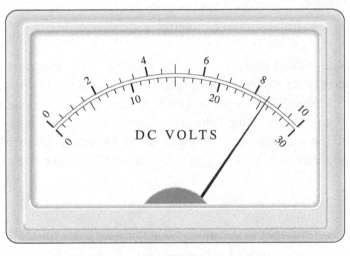

Figure 2–5

The top scale has a full-scale value of 10 V. This scale should be read if the 10 V range is selected. It is also used for any range which is a multiple of 10. For example, assume the meter shown has a 1.0 V range that has been selected. The user inserts a decimal and reads the top scale as 1.0 V full scale. The primary divisions are 0.2 V, and the secondary divisions are equal to 0.05 V. The reading on the meter is then interpreted as 0.85 V.

What is the meter reading if the range selected is the 30 V range? _____

34

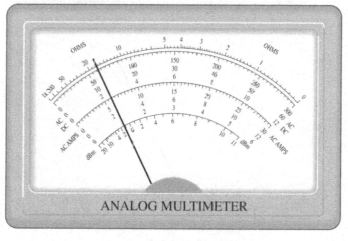

Figure 2–6

4.   VOMs and some instruments contain meters that can be used for more than one function. These scales are called *complex* scales. To read a complex scale the user chooses the appropriate scale based on the function *and* the range selected. Figure 2–6 shows a complex scale from a VOM.

If the function selected is resistance, then the top scale is selected. Before using a VOM on the resistance scale, the meter is adjusted for a zero reading with no resistance. This scale is nonlinear. Notice that the secondary divisions change values across the scale. To determine the reading, the primary divisions on each side of the pointer are noted. The secondary divisions can then be assigned values by counting the number of secondary divisions between the primary marks.

For the meter in Figure 2–6, assume the OHMS function is selected, and the range selected is ×10 ohms. What does the meter indicate for a resistance? _____

Assume the meter is on the DC VOLTS function and the range selected is 12 V. What does the meter indicate for voltage? _____

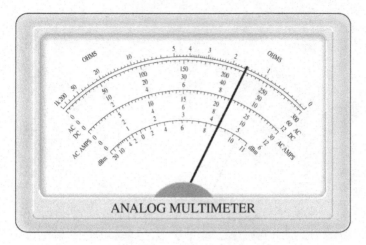

ANALOG MULTIMETER

Figure 2–7

5. For the meter in Figure 2–7, assume the OHMS function is selected and the range selected is ×100. What does the meter read for resistance? _____

    Assume the meter in Figure 2–7 is on the AC VOLTS function and the range selected is 30 V. What does the meter indicate for voltage? _____

    The decibel scale (the lowest scale) is used in 600 Ω systems (such as many audio systems). It is a logarithmic scale in which the 0 dBm reference represents a power dissipation of 1 mW in a 600 Ω system (0.775 V across 600 Ω). For the meter in Figure 2–7, what is the reading in dBm? _____

6. Look at the meter on the power supply at your lab station. Some power supplies have meters that monitor either voltage or current. There may be more than one range or several supplies built into the same chassis, so the meter may have multiple or complex scales.

    Is the meter used for more than one function? _____ If so, what determines which function is monitored? _____

    Does the meter have multiple scales? _____ complex scales? _____

    What is the smallest primary voltage division? _____ The smallest secondary voltage division? _____

7.	Review the controls for the power supply at your lab station. The operator's manual is a good resource if you are not sure of the purpose of a control. Describe the features of your supply: (multiple outputs, current limiting, tracking, etc.)

8.	In this step, you will set the power supply for a specific voltage and measure that voltage with your laboratory meter. Review the operator's manual for the DMM (or VOM) at your lab station. Review each control on the meter. Then select +DC and VOLTS on the DMM. If your DMM is not autoranging, select a range that will measure +5.0 V dc. The best choice is a range which is the *smallest* range that is larger than +5.0 V. Connect the test leads together and verify that the reading is zero. (Note: A digital meter may have a small digit in the least significant place.)

9.	Turn on the power supply at your station and use the meter on the supply to set the output to +5.0 V. Then use the DMM to confirm that the setting is correct.

	Reading on the power supply meter = _____ Reading on the DMM = _____

10.	Set the output to +12.0 V and measure the output.

	Reading on the power supply meter = _____ Reading on the DMM = _____

11.	Set the power supply to the minimum setting and measure the output.

	Reading on the power supply meter = _____ Reading on the DMM = _____

CONCLUSION

EVALUATION AND REVIEW QUESTIONS

1.	Compare the precision of the power supply voltmeter with the DMM or VOM at your lab station. Does one meter have an advantage for measuring 5.0 V? Explain your answer.

2.   What is meant by an autoranging meter? What type is at your lab station?

3.   What is the difference between a multiple scale and a complex scale?

4.   What is the difference between a linear scale and a nonlinear scale?

5.   Assume a scale has four secondary marks between the primary marks numbered 3.0 and 4.0. If the pointer is on the first secondary mark, what is the reading on the meter?

6.   List the three basic measurements that can be made with a VOM or a DMM.

FOR FURTHER INVESTIGATION
The **sensitivity** of a panel meter is a number that describes how much current is required to obtain full-scale deflection from the meter. Meter sensitivity is easily determined with a meter calibrator. If you have a meter calibrator available, go over the operator's manual and learn how to measure the full-scale current in an inexpensive panel meter. Then obtain a small panel meter and measure its sensitivity. Summarize your results.

3 Measurement of Resistance

READING
Text, Sections 2–1 through 2–5

OBJECTIVES
After performing this experiment, you will be able to:
1. Determine the listed value of a resistor using the resistor color code.
2. Use the DMM (or VOM) to measure the value of a resistor.
3. Determine the percent difference between the measured and listed values of a resistor.
4. Measure the resistance of a potentiometer and explain its operation.

MATERIALS NEEDED
Resistors: Ten assorted values
One potentiometer (any value)

SUMMARY OF THEORY
Resistance is the opposition a substance offers to current. The unit for resistance is the *ohm,* symbolized with the Greek letter capital omega (Ω). A resistor is a component designed to have a specific resistance and wattage rating. Resistors limit current but, in doing so, produce heat. The physical size of a resistor is related to its ability to dissipate heat, *not* to its resistance. A physically large resistor can dissipate more heat than a small resistor, hence the larger one would have a higher wattage rating than the smaller one.

Resistors are either fixed (constant resistance) or variable. Fixed resistors are usually color coded with a four-band code that indicates the specific resistance and tolerance. Each color stands for a number, as described in the text and reprinted in Table 3–1 for convenience. Figure 3–1 shows how to read the resistance and tolerance of a four-band resistor.

The resistance of resistors is measured using a DMM or VOM, as described in Experiment 2. If you are using a VOM, the zero reading should be checked whenever you change ranges on the meter by

Table 3–1

	Digit	Color
	0	Black
	1	Brown
	2	Red
	3	Orange
Resistance value,	4	Yellow
first three bands	5	Green
	6	Blue
	7	Violet
	8	Gray
	9	White
Tolerance, fourth band	5%	Gold
	10%	Silver
	20%	No band

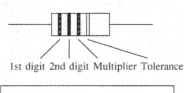

1st digit 2nd digit Multiplier Tolerance

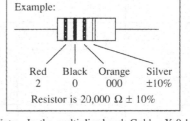

Example:

Red   Black   Orange   Silver
2   0   000   ±10%

Resistor is 20,000 Ω ± 10%

Note: In the multiplier band, Gold = X 0.1
Silver = X 0.01

Figure 3–1

touching the test leads together. If you are using a nonautoranging DMM, a suitable range needs to be selected. Resistance normally should not be measured in a circuit as other resistors in the circuit will affect the reading. The resistor to be measured is removed from the circuit, and the test leads are connected across the resistance. The resistor under test should not be held between the fingers as body resistance can affect the reading, particularly with high-value resistors. (It is okay to hold one end of the resistor under test.)

The most common form of variable resistor is the potentiometer. The potentiometer is a three-terminal device with the outer terminals having a fixed resistance between them and the center terminal connected to a moving wiper. The moving wiper is connected to a shaft that is used to vary the resistance between it and the outer terminals. Potentiometers are commonly found in applications such as volume controls.

Another type of variable resistor is the rheostat. A rheostat consists of two terminals. The control varies the resistance between the two terminals. A potentiometer can be connected as a rheostat by connecting the moving wiper and one of the outer terminals.

PROCEDURE

1. Obtain 10 four-band fixed resistors. Record the colors of each resistor in Table 3–2. Use the resistor color code to determine the color-code resistance of each resistor. Then measure the resistance of each resistor and record the measured value in Table 3–2. The first line has been completed as an example.

2. Compute the percent difference between the measured and color-coded values using the equation:

$$\% \text{ difference} = \frac{|R_{measured} - R_{color\,code}|}{R_{color\,code}} \times 100$$

The percent difference is shown as an absolute (positive) value for all resistors. Complete Table 3–2.

Table 3–2

Resistor	Color of Band				Color-Code Value	Measured Value	% Difference
	1st	2nd	3rd	4th			
0	brown	green	red	silver	1.5 kΩ ± 10%	1.46 kΩ	2.7%
1							
2							
3							
4							
5							
6							
7							
8							
9							
10							

3. Obtain a potentiometer. Number the terminals 1, 2, and 3 as illustrated in Figure 3–2.

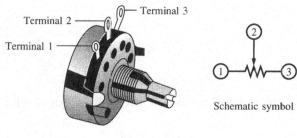

Schematic symbol

Figure 3–2

Measure and record the resistance between terminals 1 and 3 of the potentiometer (the outside terminals). $R_{1,3}$ = _____

Vary the potentiometer's shaft and monitor the resistance between terminals 1 and 3. Does the resistance change? _____ Explain. _____

4. Turn the potentiometer completely counterclockwise (CCW). Measure the resistance between terminals 1 and 2. Then measure the resistance between terminals 2 and 3. Record the measured resistance in Table 3–3. Compute the sum of the two readings and enter it into Table 3–3.

5. Turn the shaft 1/3 turn clockwise (CW) and repeat the measurements in step 4.

6. Turn the shaft 2/3 turn CW and repeat the measurements in step 4. What did you find about the sum of the resistance in steps 4, 5, and 6?

Table 3–3

Step	Shaft Position	Resistance Measured Between		Sum of Resistance Readings
		Terminals 1–2	Terminals 2–3	
4	CCW			
5	⅓ CW			
6	⅔ CW			

CONCLUSION

41

EVALUATION AND REVIEW QUESTIONS

1.   Predict the resistance between terminals 1–2 and 2–3 for the potentiometer if the shaft is rotated fully CW.

2.   (a)   Are any of the resistors measured in Table 3–2 out of tolerance? _____

        (b)   You suspect that the percent difference between color-code and measured values could be due to error in the meter. How could you find out if you are correct?

3.   Determine the resistor color code for the following resistors. The tolerance is 10%.
        (a)   12 Ω _____
        (b)   6.8 kΩ _____
        (c)   910 Ω _____
        (d)   4.7 MΩ _____
        (e)   1.0 Ω _____

4.   Determine the expected value and tolerance for resistors with the following color codes:
        (a)   red-red-black-gold _____
        (b)   violet-green-brown-silver _____
        (c)   green-brown-brown-gold _____
        (d)   white-brown-gold-gold _____
        (e)   gray-red-yellow-silver _____

5.   A resistor is color coded: red-violet-orange-gold.
        (a)   What is the largest value the resistor can be and still be in tolerance? _____

        (b)   What is the smallest value? _____

6.   What is the difference between a potentiometer and a rheostat?

FOR FURTHER INVESTIGATION

Obtain 20 resistors marked with the same color code on each resistor. Carefully measure each resistor. Determine the average value of resistance. Then find the deviation of each resistor from the average by computing the amount each value differs from the average. The average deviation can then be found by computing the average of the deviations. Summarize your findings. Are any resistors outside the marked tolerance? Is there any evidence that all resistors are either higher or lower than their marked values? What do differing values of resistors tell you about actual measurements as opposed to calculated or color-code values?

4 Voltage Measurement and Reference Ground

Name _____
Date _____
Class _____

READING
Text, Sections 2–6 and 2–7. (Optional: See Section 4–9 for additional information on reference ground.)

OBJECTIVES
After performing this experiment, you will be able to:
1.   Connect a circuit from a schematic diagram.
2.   Use voltages measured with respect to ground to compute the voltage drop across a resistor.
3.   Explain the meaning of reference ground and subscripts used in voltage definitions.

MATERIALS NEEDED
Resistors:
      One 330 Ω, one 680 Ω, and 1.0 kΩ

SUMMARY OF THEORY
Energy is required to move a charge from a point of lower potential to one of higher potential. Voltage is a measure of this energy per charge. Energy is given up when a charge moves from a point of higher potential to one of lower potential.

    Voltage is always measured with respect to some point in the circuit. For this reason, only potential *differences* have meaning. We can refer to the voltage *across* a component, in which case the reference is one side of the component. Alternatively, we can refer to the voltage at some point in the circuit. In this case the reference point is assumed to be "ground." Circuit ground is usually called *reference ground* to differentiate it from the potential of the earth, which is called *earth ground*. Various symbols for ground are shown in Figure 2–46 of the text. This manual uses the symbol in Figure 2–46(a) throughout.

    An analogy can clarify the meaning of reference ground. Assume a building has two floors below ground level. The floors in the building could be numbered from the ground floor, by numbering the lower floors with negative numbers. Alternatively, the reference for numbering the floors could be made the lowest floor in the basement. Then all floors would have a positive floor number. The choice of the numbering system does not change the height of the building, but it does change each floor number. Likewise, the ground reference is used in circuits as a point of reference for voltage measurements. The circuit is not changed by the ground reference chosen.

Figure 4–1 illustrates the same circuit with two different ground reference points. The circuit in Figure 4–1(a) has as its reference point **B.** Positive and negative voltages are shown. If the reference point is moved to point **C,** the circuit voltages are all positive, as shown in Figure 4–1(b). Voltage is always measured between two points. To define the two points, subscripts are used. The voltage difference (or simply voltage) between points **A** and **B** is written as V_{AB} where the second letter in the subscript identifies the reference point. If a single subscripted letter is shown, the voltage is defined between the lettered point and the circuit's reference ground.

For *both* circuits: $V_{AB} = 4$ V $V_{BC} = 8$ V $V_{AC} = 12$ V

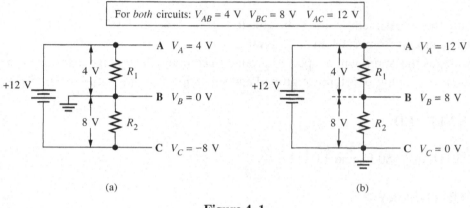

(a)   (b)

Figure 4–1

PROCEDURE

1.   Measure three resistors with the listed values given in Table 4–1. Record the measured values in Table 4–1. You should always use the measured value in experimental work.

2.   Construct the circuit shown in Figure 4–2. An example of how a series circuit like this can be built on a protoboard (solderless breadboard) is shown in the Introduction to the Student section, Figure I–2. Set the power supply to +10 V. Measure the voltage across each resistor in the circuit. Enter the measured values in Table 4–2.

Table 4–1

Component	Listed Value	Measured Value
R_1	330 Ω	
R_2	680 Ω	
R_3	1.0 kΩ	

Table 4–2

	Measured Value
V_S	
V_{AB}	
V_{BC}	
V_{CD}	

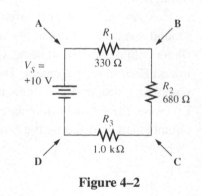

Figure 4–2

44

3.   Assign point **D** as the reference ground. Measure the voltage at points **A, B,** and **C** with respect to point **D.** The voltage readings are made with the reference probe connected to point **D.** Enter the measured values in Table 4–3. Then use the measured voltages to compute the voltage differences V_{AB}, V_{BC}, and V_{CD}.

Table 4–3

	Measured Voltage	Voltage Difference Calculation
V_A		
		$V_{AB} = V_A - V_B =$
V_B		
		$V_{BC} = V_B - V_C =$
V_C		
		$V_{CD} = V_C - V_D =$
V_D	0.0 V (ref)	

4.   Now measure the voltages in the circuit with respect to point **C.** The circuit is *not changed.* Only the reference point changes. Move the reference probe of the voltmeter to point **C.** This point will now represent ground. The voltage at point **D** now has a negative value. Enter the measured voltages in Table 4–4. Compute the voltage differences as before and enter them in Table 4–4.

Table 4–4

	Measured Voltage	Voltage Difference Calculation
V_A		
		$V_{AB} = V_A - V_B =$
V_B		
		$V_{BC} = V_B - V_C =$
V_C	0.0 V (ref)	
		$V_{CD} = V_C - V_D =$
V_D		

5.   Move the circuit reference point to point **B.** Again, there is no change to the circuit other than the reference ground. Repeat the measurements of the voltages with respect to circuit ground. Compute the voltage differences and enter the data in Table 4–5.

Table 4–5

	Measured Voltage	Voltage Difference Calculation
V_A		
		$V_{AB} = V_A - V_B =$
V_B	0.0 V (ref)	
		$V_{BC} = V_B - V_C =$
V_C		
		$V_{CD} = V_C - V_D =$
V_D		

6.   Now make point **A** the reference point and repeat the measurements. Enter the data in Table 4–6.

Table 4–6

	Measured Voltage	Voltage Difference Calculation
V_A	0.0 V (ref)	
V_B		$V_{AB} = V_A - V_B =$
V_C		$V_{BC} = V_B - V_C =$
V_D		$V_{CD} = V_C - V_D =$

CONCLUSION

EVALUATION AND REVIEW QUESTIONS

1.   Compare the *voltage difference calculation* in Table 4–3 through Table 4–6. Does the circuit's reference point have any effect on the voltage differences across any of the resistors? Explain your answer.

2.   Define the term *reference ground.*

3.   If you measured V_{AB} as 12.0 V, what is the V_{BA}? _____

4.   Assume $V_M = -220$ V and $V_N = -150$ V. What is V_{MN}? _____

5.   If a test point in a circuit is marked +5.0 V and a second test point is marked −3.3 V, what voltage reading would you expect on a voltmeter connected between the two test points? Assume the reference lead on the meter is at the lowest potential. _____

FOR FURTHER INVESTIGATION

Warning: The power supplies used in this procedure must have floating output terminals. If you are not sure, check with your instructor before proceeding with this investigation.

Replace the +10 V supply used in this experiment with two +5 V supplies in series. Attach the +5 V output of one supply to the common of the second supply. Call this point the reference ground for the circuit. Measure the voltages throughout the circuit. Summarize your results.

Application Assignment 2

Name _____
Date _____
Class _____

REFERENCE

Text, Chapter 2; Application Assignment: Putting Your Knowledge to Work. (*Note:* The Application Assignment worksheets correlate to the Application Assignments in the text.)

Step 1 Circuit choice and reason the rejected circuit will not meet requirements:

_____

_____

Purpose of components:

Battery _____

Rheostat _____

Fuse _____

Rotary switch _____

Bulbs _____

Pushbutton switches _____

Step 2 Materials list: (The first two items are completed as an example.)

Item	Description	Quantity
1	Lantern battery, 12 V	1
2	25 W rheostat, 10 W	1

Step 3   Component list and connections: (The first two items are completed as an example.)

From	To
Ground (node 1)	Battery negative side
Battery positive side (node 2)	Rheostat lower end and wiper

Step 4   Required current rating for fuse:_____

Reason_____

Step 5   The rotary switch selects question 1, but the light does not turn on. Other lights work when they are selected.

Possible trouble(s)_____

None of the lights will turn on.

Possible trouble(s)_____

The correct answer light never is on, no matter what combination of question or answer is selected.

Possible trouble(s)_____

All lights are too dim when selected.

Possible trouble(s)_____

48

Checkup 2

Name _____
Date _____
Class _____

REFERENCE
Text, Chapter 2; Lab manual, Experiments 2, 3, and 4

1. A material that is characterized by four valence electrons in its atomic structure is called:
   (a) a conductor   (b) an insulator   (c) a semiconductor

2. The basic particle of matter that carries a negative electrical charge is the:
   (a) atom   (b) electron   (c) proton   (d) neutron

3. The unit of electrical charge is the:
   (a) ampere   (b) coulomb   (c) joule   (d) volt

4. One coulomb passing a point in one second is defined as one:
   (a) ampere   (b) watt   (c) joule   (d) volt

5. A joule per coulomb is a measure of:
   (a) resistance   (b) power   (c) voltage   (d) current

6. The unit of resistance is named in honor of:
   (a) Joule   (b) Watt   (c) Ampere   (d) Ohm

7. A resistor used to control current in a circuit is called a:
   (a) circuit breaker   (b) rheostat   (c) potentiometer   (d) choke

8. The purpose of the third band of a four-band resistor is:
   (a) multiplier   (b) tolerance   (c) reliability   (d) temperature

9. An instrument used for measuring resistance is:
   (a) an ohmmeter   (b) a voltmeter   (c) an oscilloscope   (d) an ammeter

10. In a circuit, a reference ground is always:
    (a) the point with the lowest potential   (b) a common point
    (c) the same as earth ground   (d) the negative side of the source

11. If you are constructing a circuit for testing, why is it a good idea to measure and record the values of resistors used?

12. A 5.6 kΩ resistor has a fourth band that is gold. What are the largest and smallest values of resistance that are within the tolerance rating for this resistor?

13. In Experiment 4, V_{AB} was positive despite the location of ground. Explain why this was true.

14. Determine the color-code value of resistance and the tolerance for each resistor:
    (a) white-brown-red-silver: _____
    (b) green-blue-green-gold: _____
    (c) brown-black-black-gold: _____
    (d) yellow-violet-orange-silver: _____
    (e) green-brown-gold-gold: _____

15. Determine the color code for each of the following resistors:
    (a) 470 kΩ ± 10% _____
    (b) 180 Ω ± 5% _____
    (c) 4.3 kΩ ± 5% _____
    (d) 1.0 Ω ± 10% _____
    (e) 2.7 MΩ ± 5% _____

16. Explain how both positive and negative voltages can exist at the same time in a circuit but with only one voltage source.

17. Assume that a circuit contains three points labeled **A, B,** and **C**. Point **A** has a potential with respect to ground of 10.2 V; point **B** has a potential of −12.4 V, and point **C** has a potential of −8.7 V. What is the potential difference between:
    (a) point **A** with respect to point **B**? _____
    (b) point **B** with respect to point **A**? _____
    (c) point **A** with respect to point **C**? _____
    (d) point **C** with respect to point **A**? _____
    (e) point **B** with respect to point **C**? _____
    (f) point **C** with respect to point **B**? _____

18. What is another word for potential difference?

5 Ohm's Law

Name _____

Date _____

Class _____

READING
Text, Sections 3–1 and 3–2

OBJECTIVES
After performing this experiment, you will be able to:
1. Measure the current-voltage curve for a resistor.
2. Construct a graph of the data from objective 1.
3. Given a graph of current-voltage for a resistor, determine the resistance.

MATERIALS NEEDED
Resistors:

      One 1.0 kΩ, one 1.5 kΩ, one 2.2 kΩ

One dc ammeter, 0–10 mA

For Further Investigation:

  One Cds photocell (Jameco 120299 or equivalent)

SUMMARY OF THEORY
The flow of electrical charge in a circuit is called *current*. Current is measured in units of *amperes,* or amps for short. The ampere is defined as one coulomb of charge moving past a point in one second. Current is symbolized by the letter *I* (for *Intensity*) and is frequently shown with an arrow to indicate the direction of flow. Conventional current is defined as the direction a positive charge would move under the influence of an electric field. When electrons move, the direction is opposite to the direction defined for conventional current. To clarify the difference, the term *electron flow* is frequently applied to current in the opposite direction of conventional current. The experiments in this lab book work equally well with either definition.

     The relationship between current and voltage is an important characteristic that defines various electronic devices. The relationship is frequently shown with a graph. Usually, the voltage is controlled (the independent variable), and the current is observed (the dependent variable). This is the basic method for this experiment, for which a series of resistors will be tested. As discussed in the Introduction to the Student, the independent variable is plotted along the *x*-axis and the dependent variable is plotted along the *y*-axis.

     Fixed resistors have a straight-line or *linear* current-voltage curve. This linear relationship illustrates the basic relationship of Ohm's law—namely, that the current is proportional to the voltage for constant resistance. Ohm's law is the most important law of electronics. It is written in equation form as:

$$I = \frac{V}{R}$$

where *I* represents current, *V* represents voltage, and *R* represents resistance.

PROCEDURE

1. Measure three resistors with listed values of 1.0 kΩ, 1.5 kΩ, and 2.2 kΩ. Record the measured values in Table 5–1.

Table 5–1

Component	Listed Value	Measured Value
R_1	1.0 kΩ	
R_2	1.5 kΩ	
R_3	2.2 kΩ	

2. Connect the circuit shown in Figure 5–1(a). Notice that the ammeter is in series with the resistor and forms a single "loop" as shown in the protoboard wiring diagram in Figure 5–1(b). The voltmeter is then connected directly across the resistor.

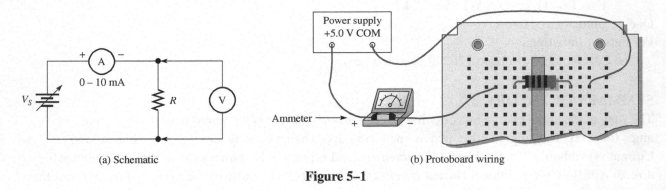

(a) Schematic   (b) Protoboard wiring

Figure 5–1

Caution!   Ammeters can be easily damaged if they are incorrectly connected. Have your instructor check your connections before applying power.

---

3. Adjust the power supply for a voltage of 2.0 V. Read the current that is through the resistor and record it in Table 5–2.

4. Adjust the power supply for 4.0 V and measure the current. Record the current in Table 5–2. Continue taking current readings for each of the voltages listed in Table 5–2.

Table 5–2 (R_1)

$V_S =$	2.0 V	4.0 V	6.0 V	8.0 V	10.0 V
$I =$					

5.   Replace R_1 with R_2 and repeat steps 3 and 4. Record the data in Table 5–3.

Table 5–3 (R_2)

$V_S =$	2.0 V	4.0 V	6.0 V	8.0 V	10.0 V
$I =$					

6.   Replace R_2 with R_3 and repeat steps 3 and 4. Record the data in Table 5–4.

Table 5–4 (R_3)

$V_S =$	2.0 V	4.0 V	6.0 V	8.0 V	10.0 V
$I =$					

7.   On Plot 5–1, graph all three *I-V* curves using the data from Tables 5–2, 5–3, and 5–4. Plot the dependent variable (current) on the *y*-axis and the independent variable (voltage) on the *x*-axis. Choose a scale for the graph that spreads the data over the entire grid.

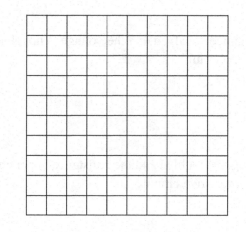

I

V

Plot 5–1

CONCLUSION

EVALUATION AND REVIEW QUESTIONS

1. The slope of a line is the change in the *y* direction divided by the change in the *x* direction. The definition for slope is illustrated in Figure 5–2. Find the slope for each resistor on Plot 5–1. Notice that the slope has units. If the change in *y* is measured in mA and the change in *x* is measured in V, the slope is mA/V = mS.

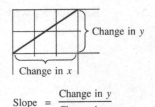

Change in *y*

Change in *x*

$$\text{Slope} = \frac{\text{Change in } y}{\text{Change in } x}$$

$$= \frac{2}{3}$$

Figure 5–2

2. What happens to the slope of the *I-V* curve for larger resistors?

3. (a) If the resistance is halved and the voltage is not changed, what will happen to the current in a resistive circuit?

   (b) If the voltage is doubled and the resistance is not changed, what will happen to the current in a resistive circuit?

4. If the current in a resistive circuit is 24 mA and the applied voltage is 48 V, what is the resistance?

5. What current is through a 10 Ω resistor with a 5.0 V applied?

FOR FURTHER INVESTIGATION

One interesting type of resistor is called a CdS cell (for Cadmium Sulfide). CdS cells are widely used as light-sensing elements in electronics. The resistance of the CdS cell decreases as the incident light increases.

In this investigation, find out if a CdS cell has an I-V curve like other resistors (a straight line) if the light is constant. Set up the experiment as in Figure 5–1, but use a CdS cell instead of a normal resistor. You will need to have a constant amount of light on the CdS cell as much as possible. Try adjusting the CdS cell to look at a light source such as a room light. You will notice that pointing it in different ways will change the current. A good starting point is to adjust it so that you have about 2 mA when the source voltage is 2.0 V. Then increase the voltage by increments of 2.0 V and record the current in Table 5–5 for each voltage setting. Plot the data in Plot 5–2 and summarize your findings.

Table 5–5 (CdS cell)

$V_S =$	2.0 V	4.0 V	6.0 V	8.0 V	10.0 V
$I =$					

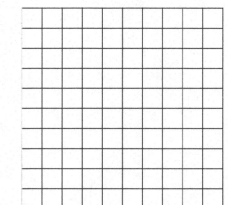

I

V

Plot 5–2

6 Power in DC Circuits

Name _____

Date _____

Class _____

READING
Text, Sections 3–3 through 3–6

OBJECTIVES
After performing this experiment, you will be able to:
1.   Determine the power in a variable resistor at various settings of resistance.
2.   Plot data for power as a function of resistance. From the plot, determine when maximum power is delivered to the variable resistor.

MATERIALS NEEDED
One 2.7 kΩ resistor
One 10 kΩ potentiometer

SUMMARY OF THEORY
When there is current through a resistor, electrical energy is converted into heat. Heat is then radiated from the resistor. The *rate* that heat is dissipated is called *power*. Power is measured in units of joules per second (J/s), which defines the unit called the watt (W). The power dissipated by a resistor is given by the power law equation:

$$P = IV$$

By applying Ohm's law to the power law equation, two more useful equations for power can be found. These are:

$$P = I^2R$$

and

$$I = \frac{V^2}{R}$$

     The three power equations given above are also known as Watt's law. In this experiment, you will determine power using the last equation. Notice that if you measure the voltage in volts (V) and the resistance in kilohms (kΩ), the power will have units of milliwatts (mW).
     The physical size of a resistor is related to the amount of heat it can dissipate. Therefore, larger resistors are rated for more power than smaller ones. Carbon composition resistors are available with standard power ratings ranging from 1/8 W to 2 W. For most typical low voltage applications (15 V or less and at least 1 kΩ of resistance), a 1/4 W resistor is satisfactory.

PROCEDURE

1. Measure the resistance of R_1. The color-code value is 2.7 kΩ. $R_1 = $ _____

2. Construct the circuit shown in Figure 6–1(a). Figure 6–1(b) shows an example of the circuit constructed on a protoboard. R_2 is a 10 kΩ potentiometer. Connect the center (variable) terminal to one of the outside terminals. Use this and the remaining terminal as a variable resistor. Adjust the potentiometer for 0.5 kΩ. (Always remove power when measuring resistance and make certain you are measuring only the potentiometer's resistance.)

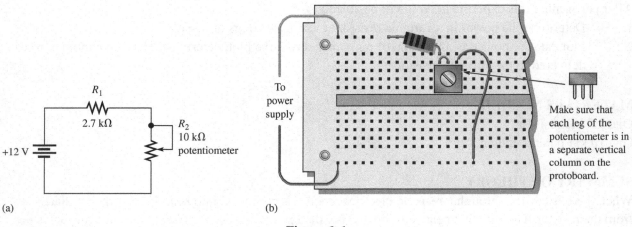

Figure 6–1

3. Use Ohm's law to compute the total current in the circuit. The total voltage is +12.0 V. The total resistance is $R_1 + R_2$. Enter the total current in Table 6–1. The first entry has been completed as an example.

Table 6–1

Variable Resistance Setting (R_2)	$I_T = \dfrac{V_T}{R_T}$	V_1 (measured)	V_2 (measured)	Power in R_2: P_2
0.5 kΩ	3.75 mA			
1.0 kΩ				
2.0 kΩ				
3.0 kΩ				
4.0 kΩ				
5.0 kΩ				
7.5 kΩ				
10.0 kΩ				

4. Measure the voltage across R_1 and the voltage across R_2. Enter the measured voltages in Table 6–1. As a check, make sure that the sum of V_1 and V_2 is equal to 12.0 V. Then compute the power in R_2 using either of the following equations:

$$P_2 = I_T R_2 \qquad \text{or} \qquad P_2 = \frac{V_2^2}{R_2}$$

Enter the computed power, in milliwatts, in Table 6–1.

5. Disconnect the power supply and set R_2 to the next value shown in Table 6–1. Reconnect the power supply and repeat the measurements made in steps 3 and 4. Continue in this manner for each of the resistance settings shown in Table 6–1.

6. Using the data in Table 6–1, graph the relationship of the power, P_2, as a function of resistance R_2 on Plot 6–1. Since resistance is the independent variable, plot it along the x-axis and plot power along the y-axis. An *implied* data point can be plotted at the origin because there can be no power dissipated in R_2 without resistance. A smooth curve can then be drawn to the origin.

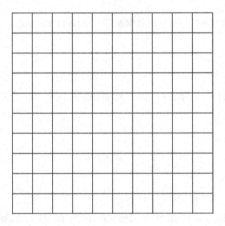

Plot 6–1

CONCLUSION

EVALUATION AND REVIEW QUESTIONS
1. Observe the graph of resistance versus power for your experiment. Compare the resistance of R_1 and R_2 when power in R_2 is a maximum.

2. What was happening to the total current in the circuit as R_2 was increasing?

3.   What was happening to the power in R_1 as the resistance of R_2 was increasing? Explain your answer.

4.   A 1.5 kΩ resistor is found to have 22.5 V across it.
        (a)   What is the current in the resistor? _____

        (b)   What is the power dissipated in the resistor? _____

        (c)   Could a 1/4 W resistor be used in this application? Explain your answer.

5.   What physical characteristic determines the power rating of a resistor?

6.   What happens to electrical energy in a resistor?

FOR FURTHER INVESTIGATION
Because it is a series circuit, the current was the same throughout for each setting of R_2. Find the current for each row in Table 6–1 by dividing the measured value of V_1 by the measured value of R_1. Plot this current as a function of R_2. On the same graph, plot V_2 as a function of R_2. What is the shape of the product of these two lines?

Application Assignment 3

Name _____

Date _____

Class _____

REFERENCE

Text, Chapter 3; Application Assignment: Putting Your Knowledge to Work

Step 1   Complete the table, showing the wattage rating for each resistor.

Resistor	Wattage Rating	Resistor	Wattage Rating
10 Ω		470 Ω	
22 Ω		1.0 kΩ	
47 Ω		2.2 kΩ	
100 Ω		4.7 kΩ	
220 Ω			

Step 2   Materials list and cost: (The first three columns are completed to save you time. Complete the last column and determine the total cost.)

Item	Description	Cost (each)	Quantity
1	1/4 W resistor	$0.08	
2	1/2 W resistor	$0.09	
3	1 W resistor	$0.09	
4	2 W resistor	$0.10	
5	5 W resistor	$0.33	
6	1 pole, 9-position rotary switch	$10.30	
7	Knob	$3.30	
8	Enclosure	$8.46	
9	Screw terminal (dual)	$0.20	
10	Binding posts	$0.60	
11	PC board	$1.78	
12	Miscellaneous	$0.50	

Total cost of project: _____

Step 3   Draw the schematic, showing resistor values and wattage ratings.

Step 4   Develop a test procedure to ensure the resistor box is working correctly.

Step 5   Describe the most likely fault for each of the following problems and how you would
                  check to verify the problem:
                  1. The ohmmeter reads infinite resistance for the 10 Ω position.

                  2. The ohmmeter reads infinite resistance for all positions of the switch.

                  3. All resistors read 10% higher than the listed value.

RELATED EXPERIMENT

MATERIALS NEEDED
One LED
One resistor to be determined

DISCUSSION
It is frequently necessary to use a resistor to limit current. A good example of this is in limiting current to
a light-emitting diode (LED), which will burn out if it has excessive current. Assume an LED requires
8 mA to function properly and drops about 2.0 V. If you want to use a 5.0 V power supply, you need to
choose a series current-limiting resistor that will drop the remaining 3.0 V. Determine the required
resistance and power rating of the current-limiting resistor. Construct the circuit and verify that your
calculated value is correct.

EXPERIMENTAL RESULTS

Checkup 3

REFERENCE
Text, Chapter 3; Lab manual, Experiments 5 and 6

1. Ohm's law states the relationship between voltage, current, and:
   (a) power   (b) energy   (c) resistance   (d) time

2. In a given dc circuit, if the voltage were doubled and the resistance halved, the new current would be:
   (a) one-fourth   (b) one-half   (c) unchanged   (d) doubled   (e) quadrupled

3. A fixed resistance is connected across a 10 V source. The current in the resistance is found to be 21.3 μA. The value of the resistance is:
   (a) 213 μΩ   (b) 213 Ω   (c) 470 Ω   (d) 0.470 MΩ

4. A blue-gray-orange-gold resistor is connected across a 25 V source. The expected current in the resistor is:
   (a) 368 μA   (b) 368 mA   (c) 1.7 μA   (d) 1.7 mA

5. A 20 mV source is connected to a 100 kΩ load. The current in the load is:
   (a) 200 nA   (b) 200 μA   (c) 5.0 μA   (d) 5.0 mA

6. The rate at which energy is used is called:
   (a) voltage   (b) frequency   (c) conductance   (d) power

7. A megawatt is the same as:
   (a) 10^{-6} W   (b) 10^{-3} W   (c) 10^3 W   (d) 10^6 W

8. Electric utility companies charge customers for:
   (a) voltage   (b) current   (c) power   (d) energy

9. The SI unit of energy is the:
   (a) joule   (b) watt   (c) ampere   (d) kilowatt-hour

10. A 1500 W resistance heater is connected to a 115 V source. The current in the heater is:
    (a) 77 mA   (b) 8.8 A   (c) 13 A   (d) 19.6 A

11. An ammeter with an internal resistance of 0.5 Ω measures a current of 10 A. What is the voltage dropped across the ammeter?

12. In Experiment 6, a fixed resistor was in series with a variable resistor. You plotted the resistance of the variable resistor as a function of the power dissipated in it.

   (a) What would you expect the graph to look like if the fixed resistor were a lower value?

   (b) What would you expect to see if the resistance of the fixed resistor were zero?

13. A 100 Ω resistor is across a 20 V source.

   (a) Determine the current in the resistor.

   (b) Compute the power dissipated in the resistor.

14. In Experiment 5, a special caution is given regarding the connection of an ammeter in a circuit. What is the proper way to connect an ammeter?

15. A 10 W bulb is designed for use in a 12 V circuit.

   (a) What current is in the bulb when it is connected to a 12 V source?

   (b) If the bulb were placed across a 6 V source, what power would it dissipate?

7 Series Circuits

READING
Text, Sections 4–1 through 4–6

OBJECTIVES
After performing this experiment, you will be able to:
1.   Use Ohm's law to find the current and voltages in a series circuit.
2.   Apply Kirchhoff's voltage law to a series circuit.

MATERIALS NEEDED
Resistors:
   One 330 Ω, one 1.0 kΩ, one 1.5 kΩ, one 2.2 kΩ
One dc ammeter, 0–10 mA

SUMMARY OF THEORY
Consider the simple circuit illustrated in Figure 7–1. The source voltage is the total current multiplied by the total resistance as given by Ohm's law. This can be stated in equation form as

$$V_S = I_T R_T$$

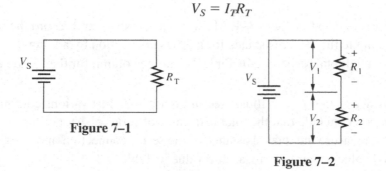

Figure 7–1

Figure 7–2

   In a series circuit, the circuit elements are connected with only one path for current. For this reason, *the current is the same throughout a series circuit.*
   Whenever we connect resistors in series, the total resistance increases. The total resistance of a series circuit is the sum of the individual resistors. Figure 7–2 illustrates a series circuit with two resistors. The total resistance is

$$R_T = R_1 + R_2$$

Substituting this equation into Ohm's law for the total circuit gives:

$$V_S = I_T(R_1 + R_2)$$

Multiplying both terms by I_T results in:

$$V_S = I_T R_1 + I_T R_2$$

65

Since the identical current, I_T, must be through each resistor, the voltage drops across the resistors can be found:

$$V_S = V_1 + V_2$$

This result illustrates that the source voltage is equal to the sum of the voltage drops across the resistors. This relationship is called Kirchhoff's voltage law, which is more precisely stated:

> The algebraic sum of all voltage rises and drops around any single closed loop in a circuit is equal to zero.

It is important to pay attention to the polarity of the voltages. Current from the source creates a voltage drop across the resistors. The voltage drop across the resistors will have an opposite polarity to the source voltage as illustrated in Figure 7–2. We may apply Kirchhoff's voltage law by using the following rules:

1. Choose an arbitrary starting point. Go either clockwise or counterclockwise from the starting point.
2. For each voltage source or load, write down the first sign you see and the magnitude of the voltage.
3. When you arrive at the starting point, equate the algebraic sum of the voltages to zero.

PROCEDURE

1. Obtain the resistors listed in Table 7–1. Measure each resistor and record the measured value in Table 7–1. Compute the total resistance for a series connection by adding the measured values. Enter the computed total resistance in Table 7–1 in the column for the listed value.

2. Connect the resistors in series as illustrated in Figure 7–3. Test various combinations of series resistors. Can you conclude that the total resistance of series resistors is the sum of the individual resistors? Then measure the total resistance of the series connection and verify that it agrees with your computed value. Enter your measured value in Table 7–1.

Table 7–1

Component	Listed Value	Measured Value
R_1	1.0 kΩ	
R_2	1.5 kΩ	
R_3	2.2 kΩ	
R_4	330 Ω	
$R_T =$		

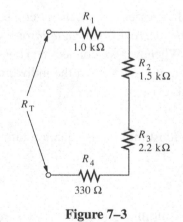

Figure 7–3

3. Complete the circuit shown in Figure 7–4. Be certain the ammeter is connected in *series,* otherwise damage to the meter may result. Before applying power, have your instructor check your circuit. Compute the current in the circuit by substituting the source voltage and the total resistance into Ohm's law. That is:

$$I_T = \frac{V_S}{R_T}$$

Record the computed current in Table 7–2. Apply power, and confirm that your computed current is within experimental uncertainty of the measured current.

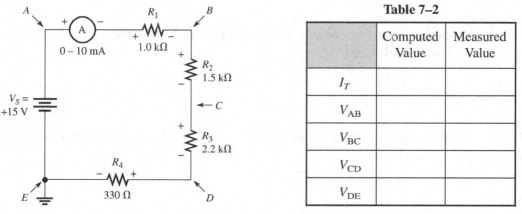

Table 7–2

	Computed Value	Measured Value
I_T		
V_{AB}		
V_{BC}		
V_{CD}		
V_{DE}		

Figure 7–4

4. In a series circuit, the same current is through all components. (Can you think of a simple proof of this?) You can use the total current measured in step 3 and Ohm's law to compute the voltage drop across each resistor. Compute V_{AB} by multiplying the total current in the circuit by the resistance between **A** and **B.** Record the results as the computed voltage in Table 7–2.

5. Repeat step 4 for the other voltages listed in Table 7–2.

6. Measure and record each of the voltages listed in Table 7–2.

7. Using the source voltage ($+15$ V) and the *measured voltage drops* listed in Table 7–2, prove that the algebraic sum of the voltages is zero. Do this by applying the rules listed in the Summary of Theory. The polarities of voltages are shown in Figure 7–4.

_____

8. Repeat step 7 by starting at a different point in the circuit and traversing the circuit in the opposite direction.

_____

9. Open the circuit at point **B.** Measure the voltage across the open circuit. Call this voltage V_{open}. Prove that Kirchhoff's voltage law is still valid for the open circuit.

_____

CONCLUSION

EVALUATION AND REVIEW QUESTIONS

1.   Why doesn't the starting point for summing the voltages around a closed loop make any difference?

2.   Kirchhoff's voltage law applies to any closed path, even one without current. How did the result of step 9 show that this is true?

3.   Based on the result you observed in step 9, what voltage would you expect in a 120 V circuit across an open (blown) fuse?

4.   Use Kirchhoff's voltage law to find V_X in Figure 7–5:

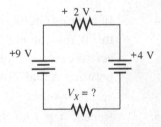

Figure 7–5

5.   A 10 Ω resistor is in series with a bulb and a 12 V source.
       (a)   If 8.0 V is across the bulb, what voltage is across the resistor? _____

       (b)   What is the current in the circuit? _____

       (c)   What is the resistance of the bulb? _____

FOR FURTHER INVESTIGATION

Resistors R_1, R_2, and R_3 used in this experiment have the same listed values as R_1, R_2, and R_3 from Experiment 5. Refer to your results of the current-voltage curve on Plot 1 of Experiment 5. Using the measured voltage in Table 7–2, find the current in the resistor based on Plot 5–1 of Experiment 5.

$I_1 =$ _____

$I_2 =$ _____

$I_3 =$ _____

What observation did you make from this about the current in a series circuit?

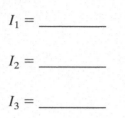

MULTISIM TROUBLESHOOTING

This experiment has four Multisim files on the website (www.prenhall.com/floyd). Three of the four files contain a simulated "fault"; one has "no fault." The file with no fault is named EXP7-4-nf. You may want to open this file to compare your results with the computer simulation. Then open each of the files with faults. Use the simulated instruments to investigate the circuit and determine the problem. The following are the filenames for circuits with troubleshooting problems for this experiment.

EXP7-4-f1
    Fault: _____

EXP7-4-f2
    Fault: _____

EXP7-4-f3
    Fault: _____

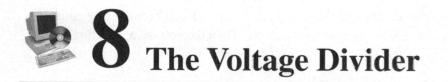

8 The Voltage Divider

Name _____
Date _____
Class _____

READING
Text, Sections 4–7 through 4–10

OBJECTIVES
After performing this experiment, you will be able to:
1. Apply the voltage divider rule to series resistive circuits.
2. Design a voltage divider to meet a specific voltage output.
3. Confirm experimentally the circuit designed in step 2.
4. Determine the range of voltages available when a variable resistor is used in a voltage divider.

MATERIALS NEEDED
Resistors:
   One 330 Ω, one 470 Ω, one 680 Ω, one 1.0 kΩ
One 1.0 kΩ potentiometer

SUMMARY OF THEORY
A voltage divider consists of two or more resistors connected in series with a voltage source. Voltage dividers are used to obtain a smaller voltage from a larger source voltage. As you saw in Experiment 7, the voltage drops in a series circuit equal the source voltage. If you have two equal resistors in series, the voltage across each will be one-half of the source voltage. The voltage has thus been divided between the two resistors. The idea can be extended to circuits with more than two resistors and with different values.

   Consider the series circuit illustrated in Figure 8–1. If the resistors are equal, the voltage across R_2 will be one-half the source voltage. But what happens if one of the resistors is larger than the other? Since both resistors must have the *same* current, Ohm's law tells us that the larger resistor must drop a larger voltage. In fact, the voltage across any resistor in a series circuit can be found by finding the *fraction* of the total resistance represented by the resistor in question. For example, if a series resistor represents one-third of the total resistance, the voltage across it will be one-third of the source voltage.

   To find the voltage across R_2, the ratio of R_2 to R_T is multiplied by the source voltage. That is:

$$V_2 = V_S \left(\frac{R_2}{R_T} \right)$$

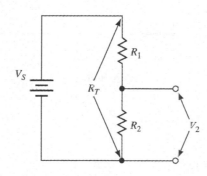

Figure 8–1

The voltage divider formula can be extended to find the voltage in a series circuit between any number of resistors. Call the resistance that is between the output terminals R_X. Then the voltage across this resistance can be written:

$$V_X = V_S \left(\frac{R_X}{R_T} \right)$$

where R_X represents the resistance between the output terminals.

This equation is a general form of the voltage divider equation. It can be stated as: "The output voltage from a voltage divider is equal to the input voltage multiplied by the ratio of the resistance between the output terminals to the total resistance." When several resistors are used, the output is generally taken with respect to the ground reference for the divider, as shown in Figure 8–2. In this case the output voltage can be found by substituting the value of R_2 and R_3 for R_X as shown.

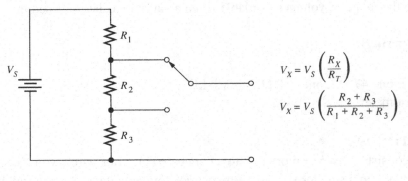

$$V_X = V_S \left(\frac{R_X}{R_T} \right)$$

$$V_X = V_S \left(\frac{R_2 + R_3}{R_1 + R_2 + R_3} \right)$$

Figure 8–2

Voltage dividers can be made to obtain variable voltages by using a potentiometer. The full range of the input voltage is available at the output, as illustrated in Figure 8–3(a). If one desires to limit the output voltage, this can be done by using fixed resistors in series as illustrated in the example shown in Figure 8–3(b).

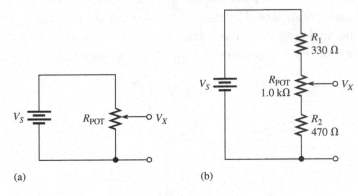

(a)   (b)

Figure 8–3

PROCEDURE

1. Obtain the resistors listed in Table 8–1. Measure each resistor and record the measured value in Table 8–1, column 3. Compute the total resistance for a series connection by adding the measured values. Enter the computed total resistance in Table 8–1.

Table 8–1

Resistor	Listed Value	Measured Value	$V_X = V_S \left(\dfrac{R_X}{R_T} \right)$	V_X (measured)
R_1	330 Ω			
R_2	470 Ω			
R_3	680 Ω			
R_4	1000 Ω			
Total			10.0 V	

2. Connect the resistors in the series circuit illustrated in Figure 8–4. With the source disconnected, measure the total resistance of the series connection and verify that it agrees with your computed value.

3. Apply the voltage divider rule to each resistor, one at a time, to compute the expected voltage across that resistor. Use the measured values of resistance and a source voltage of +10 V. Record the computed voltages (V_X) in Table 8–1, column 4.

4. Turn on the power and measure the voltage across each resistor. Record the measured voltage drops in Table 8–1, column 5. Your measured voltages should agree with your computed values.

5. Observe the voltages measured in step 4. In the space provided, draw the voltage divider, showing how you could obtain an output of +6.8 V.

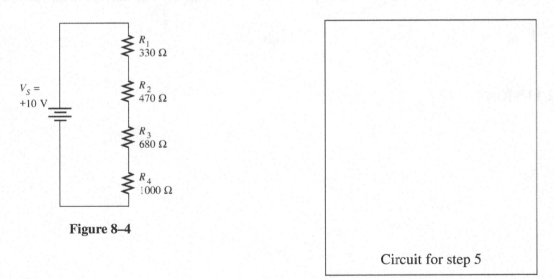

Figure 8–4

Circuit for step 5

6. Using the 330 Ω, 680 Ω, and 1.0 kΩ resistors, design a voltage divider with a +5.0 V output from a source voltage of +10 V. Draw your design in the space provided below.

7. Construct the circuit you designed and measure the actual output voltage. Indicate the measured value on your drawing.

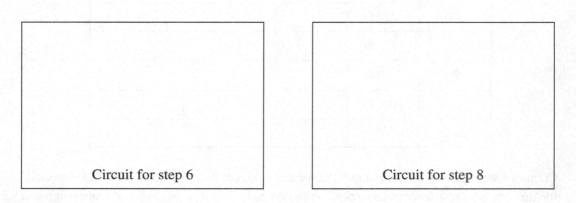

Circuit for step 6   Circuit for step 8

8. Use two of the resistors from this experiment to design a divider with a +10 V input and a +7.5 V output. Draw your design in the space provided.

9. The circuit shown in Figure 8–3(b) uses a 1.0 kΩ potentiometer and R_1 and R_2 to limit the range of voltages. Assume V_S is +10 V. Use the voltage divider formula to compute the minimum and maximum voltages available from this circuit:

$V_{MIN} =$ $V_{MAX} =$

10. Construct the circuit computed in step 9. Measure the minimum and maximum output voltages:

$V_{MIN} =$ $V_{MAX} =$

CONCLUSION

EVALUATION AND REVIEW QUESTIONS

1. (a)   If all the resistors in Figure 8–4 were 10 times larger than the specified values, what would happen to the output voltage?

    (b)   What would happen to the power dissipated in the voltage divider?

2. Refer to Figure 8–3(b). Assume V_S is 10.0 V.
    (a)   If R_1 is open, what is the output voltage? _____

    (b)   If R_2 is open, what is the output voltage? _____

3. If a student used a potentiometer in the circuit of Figure 8–3(b) that was 10 kΩ instead of 1.0 kΩ, what would happen to the range of output voltages?

4. For the circuit in Figure 8–5, compute the output voltage for each position of the switch:
    V_A _____
    V_B _____
    V_C _____
    V_D _____

5. Compute the minimum and maximum voltage available from the circuit shown in Figure 8–6:

    V_{MIN} = _____ V_{MAX} = _____

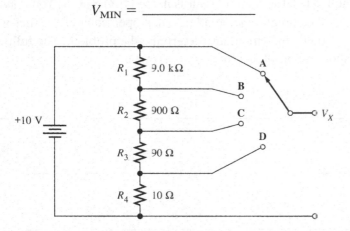

Figure 8–5

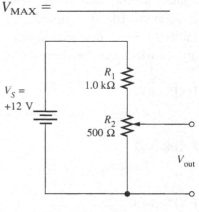

Figure 8–6

FOR FURTHER INVESTIGATION

The voltage dividers in this experiment were *unloaded*—that is, they were not required to furnish current to a load. If a load is put on the output, then current is supplied to the load and the output voltage of the divider changes. Investigate this effect by placing some load resistors on the voltage divider from this experiment (Figure 8–4). What size load resistor causes a 10% or less effect? Does the size of the resistors in the divider string affect your results? Why would you choose one set of resistors over another? Summarize your findings in a short laboratory report.

MULTISIM TROUBLESHOOTING

This experiment has four Multisim files on the website (www.prenhall.com/floyd). Three of the four files contain a simulated "fault"; one has "no fault". The file with no fault is named EXP8-3-nf. You may want to open this file to compare your results with the computer simulation. Then open each of the files with faults. Use the simulated instruments to investigate the circuit and determine the problem. The following are the filenames for circuits with troubleshooting problems for this experiment.

EXP8-3-f1

    Fault: _____

EXP8-3-f2

    Fault: _____

EXP8-3-f3

    Fault: _____

Application Assignment 4

Name _____
Date _____
Class _____

REFERENCE

Text, Chapter 4; Application Assignment: Putting Your Knowledge to Work

Step 1   Draw the schematic of the circuit.

Step 2   Determine the voltages.

	Specified (5%)	Computed
Pin 1:	0.0 V	
Pin 2:	2.7 V	
Pin 3:	12.0 V	
Pin 4:	10.4 V	
Pin 5:	8.0 V	
Pin 6:	7.3 V	
Pin 7:	6.0 V	

Step 3   Modify the existing circuit if necessary. Draw the schematic.

Step 4   Determine the life of the 6.5 Ah battery for your circuit.

Step 5   Step-by-step test procedure:

Step 6   Troubleshooting:
                Fault 1 (no voltage at any pin): _____
                Fault 2 (12 V at pins 3 and 4; all others have 0 V): _____
                Fault 3 (12 V at all pins except 0 V at pin 1): _____
                Fault 4 (12 V at pin 6 and 0 V at pin 7): _____
                Fault 5 (3.3 V at pin 2): _____

RELATED EXPERIMENT

MATERIALS NEEDED
Resistors:
        One 3.3 kΩ, one 6.8 kΩ, two 10 kΩ

DISCUSSION
Voltage dividers are commonly used to set up reference voltages. For example, a logic voltage can be compared to a specified threshold level to see if it is above or below the threshold. The voltage divider circuit shown in Figure AA–4–1 provides both positive and negative reference voltages. TTL (transistor-transistor logic) uses positive voltages, whereas ECL (emitter-coupled logic) uses negative voltages. The voltages required are shown and can be obtained from a single adjustable power supply. Use the resistors listed in the materials list to design a voltage divider that produces the voltages shown. Each voltage should be within 5% of the required voltage. Set up your circuit, and measure the voltages with respect to ground. Summarize your results in a short laboratory report.

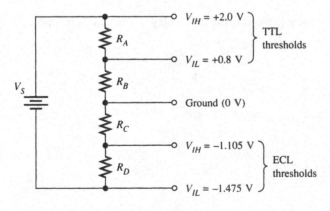

Figure AA–4–1

78

Checkup 4

REFERENCE

Text, Chapter 4; Lab manual, Experiments 7 and 8

1. In a series circuit, all components have the same:
   (a) voltage drop   (b) power   (c) resistance   (d) current

2. A 50 V power supply is connected to five 1.0 kΩ series resistors. The current in each resistor is:
   (a) 0.1 mA   (b) 10 mA   (c) 50 mA   (d) 250 mA

3. Three equal value resistors are connected in series. If the total resistance is 10 kΩ, what is the value of each resistor?
   (a) 3.3 kΩ   (b) 10 kΩ   (c) 20 kΩ   (d) 30 kΩ

4. The sum of the *IR* drops in a series circuit is:
   (a) smaller than the applied voltage   (b) equal to the largest of the *IR* drops
   (c) greater than the applied voltage   (d) equal to the applied voltage

5. A voltage source of 10 V is connected to two series resistors. The voltage across the first resistor is found to be 8.0 V. The voltage across the second resistor is:
   (a) 2.0 V   (b) 8.0 V   (c) 10 V   (d) 18 V

6. A 10 V supply is available to operate a dc motor that requires 6.0 V at 0.25 A. A series dropping resistor is needed to drop the voltage to the required level for the motor. The resistance should be:
   (a) 1.5 Ω   (b) 16 Ω   (c) 24 Ω   (d) 40 Ω

7. Two resistors are connected in series. The first resistor is found to have 5.0 V across it, and the second resistor is found to have 10 V across it. Which resistor has the greatest resistance?
   (a) the first   (b) the second   (c) neither   (d) cannot be determined

8. Two resistors are connected in series. The first resistor has 5.0 V across it, and the second resistor has 10 V across it. Which resistor dissipates the greatest power?
   (a) the first   (b) the second   (c) neither   (d) cannot be determined

9. A 75 W bulb is designed to operate from a 120 V source. If two 75 W bulbs are connected in series with a 120 V source, the total power dissipated by both bulbs is:
   (a) 37.5 W   (b) 75 W   (c) 150 W   (d) 300 W

10. A 75 W bulb is designed to operate from a 120 V source. If two 75 W bulbs are connected in series with a 240 V source, the total power dissipated by both bulbs is:
    (a) 37.5 W   (b) 75 W   (c) 150 W   (d) 300 W

11.   Find the total resistance of the series combination of a 1.20 MΩ resistor, a 620 kΩ resistor, and a 150 kΩ resistor.

12.   Assume you need a 36 V source but have only three 12 V batteries available. Draw the connection of the batteries to provide the required voltage.

13.   Assume a faulty series circuit has no current due to an open circuit. How would you use a voltmeter to locate the open circuit?

14.   A series circuit consists of three 50 Ω resistors, each rated for 250 mW.
        (a)   What is the largest voltage that can be applied before exceeding the power rating of any resistor?

        (b)   How much current is in the circuit at this voltage?

15.   The total resistance of a series circuit is 2.2 kΩ. What fraction of the input voltage will appear across a 100 Ω resistor?

16.   An 8 Ω series limiting resistor is used to limit the current in a bulb to 0.375 A with 12 V applied.
        (a)   Determine the resistance of the bulb.

        (b)   If the voltage source is increased to 15 V, what additional series resistance will limit the current to the same 0.375 A?

17.   In Experiment 7 (Series Circuits), you used resistors that ranged in value from 330 Ω to 2.2 kΩ. How would your results have changed if all of the resistors were 20% larger than called for?

9 Parallel Circuits

Name _____
Date _____
Class _____

READING
Text, Sections 5–1 through 5–8

OBJECTIVES
After performing this experiment, you will be able to:
1.   Demonstrate that the total resistance in a parallel circuit decreases as resistors are added.
2.   Compute and measure resistance and currents in parallel circuits.
3.   Explain how to troubleshoot parallel circuits.

MATERIALS NEEDED
Resistors:
        One 3.3 kΩ, one 4.7 kΩ, one 6.8 kΩ, one 10 kΩ
One dc ammeter, 0–10 mA

SUMMARY OF THEORY
A *parallel* circuit is one in which there is more than one path for current. Parallel circuits can be thought of as two parallel lines, representing conductors, with a voltage source and components connected between the lines. This idea is illustrated in Figure 9–1. The source voltage appears across each component. Each path for current is called a *branch*. The current in any branch is dependent only on the resistance of that branch and the source voltage.

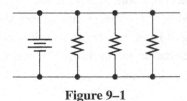

Figure 9–1

As more branches are added to a parallel circuit, the total resistance decreases. This is easy to see if you consider each added path in terms of conductance. Recall that conductance is the reciprocal of resistance. As parallel branches are added, new paths are provided for current, increasing the conductance. There is more total current in the circuit. If the total current in a circuit increases, with no change in source voltage, the total resistance must decrease according to Ohm's law. The total conductance of a parallel circuit is the sum of the individual conductances. This can be written:

$$G_T = G_1 + G_2 + G_3 + \ldots + G_n$$

By substituting the definition for resistance into the formula for conductance, the reciprocal formula for resistance in parallel circuits is obtained. It is:

$$\frac{1}{R_T} = \frac{1}{R_1} + \frac{1}{R_2} + \frac{1}{R_3} + \ldots + \frac{1}{R_n}$$

In parallel circuits, there are junctions where two or more components are connected. Figure 9–2 shows a circuit junction labeled A. Since electrical charge cannot accumulate at a point, the current into the junction must be equal to the current from the junction. In this case, $I_1 + I_2$ is equal to $I_3 + I_4$. This idea is expressed in Kirchhoff's current law, which is stated:

> The sum of the currents entering a circuit junction is equal to the sum of the currents leaving the junction.

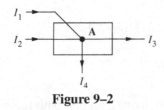

Figure 9–2

One important idea can be seen by applying Kirchhoff's current law to a point next to the source voltage. The current leaving the source must be equal to the sum of the individual branch currents. While Kirchhoff's voltage law is developed in the study of series circuits, and the current law is developed in the study of parallel circuits, both laws are applicable to any circuit.

In Experiment 8, you observed how a series circuit causes voltage to be divided between the various resistances. In parallel circuits, it is the *current* that is divided between the resistances. Keep in mind that the larger the resistance, the smaller the current. The general current divider rule can be written:

$$I_X = \left(\frac{R_T}{R_X} \right) I_T$$

Notice that the fraction R_T/R_X is always less than 1.0 and represents the fraction of the total current in R_X. This equation can be simplified for the special case of exactly two resistors. The special two-resistor current divider is written:

$$I_1 = \left(\frac{R_2}{R_1 + R_2} \right) I_T \qquad I_2 = \left(\frac{R_1}{R_1 + R_2} \right) I_T$$

PROCEDURE

1. Obtain the resistors listed in Table 9–1. Measure and record the value of each resistor.

Table 9–1

Component	Listed Value	Measured Value
R_1	3.3 kΩ	
R_2	4.7 kΩ	
R_3	6.8 kΩ	
R_4	10.0 kΩ	

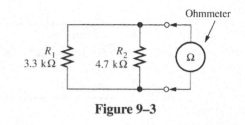

Figure 9–3

2. In Table 9–2 you will tabulate the total resistance as resistors are added in parallel. (Parallel connections are indicated with two parallel lines shown between the resistors.) Enter the measured value of R_1 in the table. Then connect R_2 in parallel with R_1 and measure the total resistance as shown in Figure 9–3. Enter the measured resistance of R_1 in parallel with R_2 in Table 9–2.

Table 9–2

	R_1	$R_1\|R_2$	$R_1\|R_2\|R_3$	$R_1\|R_2\|R_3\|R_4$
R_T (measured)				
I_T (measured)				

3. Add R_3 in parallel with R_1 and R_2. Measure the parallel resistance of all three resistors. Then add R_4 in parallel with the other three resistors and repeat the measurement. Record your results in Table 9–2.

4. Complete the parallel circuit by adding the voltage source and the ammeter as shown in Figure 9–4. Be certain that the ammeter is connected in series with the voltage source as shown. If you are not sure, have your instructor check your circuit. Measure the total current in the circuit and record it in Table 9–2.

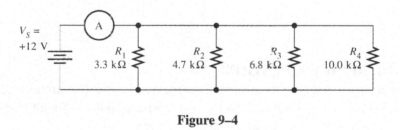

Figure 9–4

5. Measure the voltage across each resistor. How does the voltage across each resistor compare to the source voltage?

Table 9–3

	$I_1 = \dfrac{V_S}{R_1}$	$I_2 = \dfrac{V_S}{R_2}$	$I_3 = \dfrac{V_S}{R_3}$	$I_4 = \dfrac{V_S}{R_4}$
I (computed)				

6.   Use Ohm's law to compute the branch current in each resistor. Use the source voltage and the measured resistances. Tabulate the computed currents in Table 9–3.

7.   Use the general current divider rule to compute the current in each branch. Use the total current and total resistance that you recorded in Table 9–2. Compare the calculation using the current divider rule with the results using Ohm's law. Show your results in Table 9–4.

Table 9–4

	$I_1 = \left(\dfrac{R_T}{R_1}\right)I_T$	$I_2 = \left(\dfrac{R_T}{R_2}\right)I_T$	$I_3 = \left(\dfrac{R_T}{R_3}\right)I_T$	$I_4 = \left(\dfrac{R_T}{R_4}\right)I_T$
I (computed)				

8.   Prove Kirchhoff's current law for the circuit by showing that the total current is equal to the sum of the branch currents:

_____

9.   Simulate a burned-out resistor by removing R_4 from the circuit. What is the new total current?

$I_T =$ _____

CONCLUSION

EVALUATION AND REVIEW QUESTIONS

1.   In step 9, you simulated an open resistor by removing it from the circuit, and you observed that the total current dropped. Explain how the open resistor could be found in this experiment from the observed change in current and the source voltage.

2.   If one of the resistors in this experiment were shorted, what would you expect to see happen? (Do not simulate this!)

3.   Three resistors are connected in parallel across a 40 V source. The values of the resistors are 620 Ω, 750 Ω, and 820 Ω.

        (a)   What should the total source current be?

        (b)   If the measured current was 118 mA, what fault could account for this?

4.   The known currents for a circuit junction are shown in Figure 9–5. What is the value and direction of the unknown current, I_4?

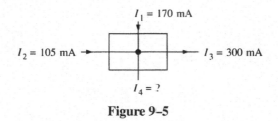

$I_1 = 170$ mA

$I_2 = 105$ mA

$I_3 = 300$ mA

$I_4 = ?$

Figure 9–5

5.   Could a shorted component in a parallel circuit cause an open to occur elsewhere? Explain.

FOR FURTHER INVESTIGATION

Kirchhoff's current law can be applied to any junction in a circuit. The currents in this circuit were I_1, I_2, I_3, I_4, and I_T. Apply Kirchhoff's current law to these currents by writing the numeric value of the current entering and leaving junction X, Y, and Z in Figure 9–6. Then verify that you computed the correct currents by measuring them with the ammeter. For each current measurement, the circuit must be broken, and the ammeter must be inserted in series with the path you are measuring. Summarize your results in a laboratory report.

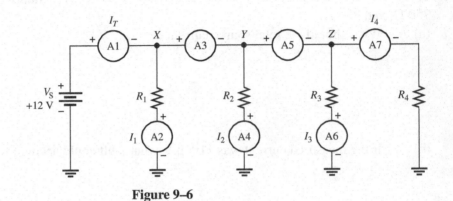

Figure 9–6

Application Assignment 5

REFERENCE

Text, Chapter 5; Application Assignment: Putting Your Knowledge to Work

■ Determine the maximum power dissipated by R_{SH} in Figure 5–52 for each range setting.

■ How much voltage is there from A to B in Figure 5–52 when the switch is set to the 2.5 A range and the current is 1 A?

■ The meter indicates 250 mA. How much does the voltage across the meter circuit from A to B change when the switch is moved from the 250 mA position to the 2.5 A position?

■ Assume the meter movement has a resistance of 4 Ω instead of 6 Ω. Specify any changes necessary in the circuit of Figure 5–52.

RELATED EXPERIMENT

MATERIALS NEEDED
Resistors:

Two 1.0 kΩ, one 1.5 kΩ, one 1.8 kΩ, one 2.2 kΩ

DISCUSSION
There are several ways of finding an open resistor in a parallel arrangement; presented here is a different method that you can investigate. The method is based on the voltage divider principle. A series 1.0 kΩ resistor is added to the parallel resistors, as shown in Figure AA–5–1. The parallel group represents an equivalent resistance in series with the 1.0 kΩ resistor. The voltage dropped across the parallel resistors will change if any resistor is open. Investigate this by connecting the circuit and measuring the voltage across the parallel group. Then, open one of the parallel resistors and measure the new voltage across the remaining group. Continue like this for each of the parallel resistors. Can you use your results to determine which resistor is open?

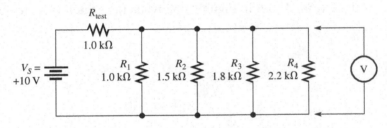

Figure AA–5–1

EXPERIMENTAL RESULTS

Checkup 5

Name _____

Date _____

Class _____

REFERENCE

Text, Chapter 5; Lab manual, Experiment 9

1.   In a parallel circuit, all components have the same:

(a) voltage drop   (b) current   (c) power   (d) resistance

2.   A 50 V power supply is connected to five 1.0 kΩ resistors connected in parallel. The total current from the source is:

(a) 0.1 mA   (b) 10 mA   (c) 50 mA   (d) 250 mA

3.   Three equal-value resistors are connected in parallel. If the total resistance is 10 kΩ, what is the value of each resistor?

(a) 3.3 kΩ   (b) 10 kΩ   (c) 20 kΩ   (d) 30 kΩ

4.   When a resistance path is added to a parallel circuit, the total resistance:

(a) decreases   (b) remains the same   (c) increases

5.   Assume a voltage of 27 V is connected across two equal parallel resistors. The current in the first resistor is 10 mA. The total resistance is:

(a) 270 Ω   (b) 741 Ω   (c) 1.35 kΩ   (d) 2.7 kΩ

6.   If one resistive branch of a parallel circuit is opened, the total current will:

(a) decrease   (b) remain the same   (c) increase

7.   Three resistors are connected in parallel. The first is 1.0 MΩ, the second is 2.0 MΩ, and the third is 10 kΩ. The total resistance is approximately:

(a) 5 kΩ   (b) 10 kΩ   (c) 1 MΩ   (d) 3 MΩ

8.   Three 75 W bulbs are connected in parallel across a 120 V line. The total power dissipated by the bulbs is:

(a) 25 W   (b) 75 W   (c) 120 W   (d) 225 W

9.   Assume an unknown resistor is in parallel with a 68 Ω resistor. The total resistance of the combination is 40.5 Ω. The resistance of the unknown resistor is:

(a) 25 Ω   (b) 34 Ω   (c) 75 Ω   (d) 100 Ω

10.   An ammeter with an internal resistance of 40 Ω and a full-scale deflection of 10 mA is needed to measure a full-scale current of 100 mA. The shunt resistor that will accomplish this has a value of:

(a) 4.0 Ω   (b) 4.44 Ω   (c) 400 Ω   (d) 444 Ω

11. In Experiment 9 (Parallel Circuits), you were asked to find the parallel resistance of a group of resistors as new ones were placed in the circuit. What was happening to the total *conductance* of the circuit as more resistors were placed in parallel? Why?

12. Explain why electrical house wiring is done with parallel circuits.

13. A 120 V source provides 30 A into a four-branch parallel circuit. The first three branch currents are 10 A, 8 A, and 5 A.
    (a)   What is the current in the fourth branch?

    (b)   What is the resistance of the fourth branch?

    (c)   What is the total resistance of the circuit?

14. Assume there is a current of 350 μA into a parallel combination of two resistors, R_1 and R_2. The resistance of R_1 is 5.6 kΩ, and the resistance of R_2 is 8.2 kΩ. Compute the current in each resistor.

15. Four 1.0 kΩ resistors are connected in parallel. The total power dissipated is 200 mW.
    (a)   What power is dissipated in each resistor?

    (b)   What is the source voltage?

16. For the parallel circuit shown in Figure C–5–1, assume the ammeter reads 1.75 mA. What is the likely cause of trouble? Justify your answer.

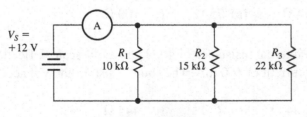

Figure C–5–1

10 Series-Parallel Combination Circuits

Name _____
Date _____
Class _____

READING
Text, Sections 6–1 through 6–4

OBJECTIVES
After performing this experiment, you will be able to:
1.   Use the concept of equivalent circuits to simplify series-parallel circuit analysis.
2.   Compute the currents and voltages in a series-parallel combination circuit and verify your computation with circuit measurements.

MATERIALS NEEDED
Resistors:
        One 2.2 kΩ, one 4.7 kΩ, one 5.6 kΩ, one 10 kΩ

SUMMARY OF THEORY
Most electronic circuits are not just series or just parallel circuits. Instead they may contain combinations of components. Many circuits can be analyzed by applying the ideas developed for series and parallel circuits to them. Remember that in a *series* circuit the same current is through all components, and that the total resistance of series resistors is the sum of the individual resistors. By contrast, in *parallel* circuits, the applied voltage is the same across all branches and the total resistance is given by the reciprocals formula.

    In this experiment, the circuit elements are connected in composite circuits containing both series and parallel combinations. The key to solving these circuits is to form equivalent circuits from the series or parallel elements. You need to recognize when circuit elements are connected in series or parallel in order to form the equivalent circuit. For example, in Figure 10–1(a) we see that the identical current must go through both R_2 and R_3. We conclude that these resistors are in series and could be replaced by an equivalent resistor equal to their sum. Figure 10–1(b) illustrates this idea. The circuit has been simplified to an equivalent parallel circuit. After finding the currents in the equivalent circuit, the results can be applied to the original circuit to complete the solution.

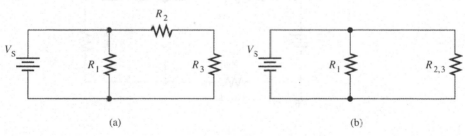

(a)   (b)

Figure 10–1

The answer to two questions will help you identify a series or parallel connection:
(1) Will the *identical* current go through both components? If the answer is yes, the components are in series. (2) Are *both ends* of one component connected directly to *both ends* of another component? If yes, the components are in parallel. The components that are in series or parallel may be replaced with an equivalent component. This process continues until the circuit is reduced to a simple series or parallel circuit. After solving the equivalent circuit, the process is reversed in order to apply the solution to the original circuit. This idea is studied in this experiment.

PROCEDURE

1.   Measure and record the actual values of the four resistors listed in Table 10–1.

Table 10–1

Component	Listed Value	Measured Value
R_1	2.2 kΩ	
R_2	4.7 kΩ	
R_3	5.6 kΩ	
R_4	10.0 kΩ	

2.   Connect the circuit shown in Figure 10–2. Then answer the following questions:

(a)   Are there any resistors for which the identical current will go through the resistors? Answer yes or no for each resistor:

R_1 _____ R_2 _____ R_3 _____ R_4 _____

(b)   Does any resistor have both ends connected directly to both ends of another resistor? Answer yes or no for each resistor:

R_1 _____ R_2 _____ R_3 _____ R_4 _____

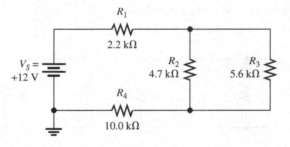

Figure 10–2

3. The answer to these questions should clarify in your mind which resistors are in series and which resistors are in parallel. You can begin solving for the currents and voltages in the circuit by replacing resistors that are either in series or in parallel with an equivalent resistor. In this case, begin by replacing R_2 and R_3 with an equivalent resistor labeled $R_{2,3}$. Draw the equivalent circuit in the space provided. Show the value of all components including $R_{2,3}$.

4. The equivalent circuit you drew in step 3 is a simple series circuit. Compute the total resistance of this equivalent circuit and enter it in the first two columns of Table 10–2. Then disconnect the power supply and measure the total resistance to confirm your calculation.

Table 10–2

| | Computed | | Measured |
	Voltage Divider	Ohm's Law	
R_T			
I_T			
V_1			
$V_{2,3}$			
V_4			
I_2			
I_3			
V_T	12.0 V	12.0 V	

5. The voltage divider rule can be applied directly to the series equivalent circuit to find the voltages across R_1, $R_{2,3}$, and R_4. Find V_1, $V_{2,3}$, and V_4 using the voltage divider rule. Tabulate the results in Table 10–2 in the Voltage Divider column.

6. Find the total current, I_T, in the circuit by substituting the total voltage and the total resistance into Ohm's law. Enter the computed total current in Table 10–2 in the Ohm's Law column.

7. In the equivalent series circuit, the total current is through R_1, $R_{2,3}$, and R_4. The voltage drop across each of these resistors can be found by applying Ohm's law to each resistor. Compute V_1, $V_{2,3}$, and V_4 using this method. Enter the voltages in Table 10–2 in the Ohm's Law column.

8. Use $V_{2,3}$ and Ohm's law to compute the current in R_2 and R_3 of the original circuit. Enter the computed current in Table 10–2. As a check, verify that the computed sum of I_2 and I_3 is equal to the computed total current.

9. Measure the voltages V_1, $V_{2,3}$, and V_4. Enter the measured values in Table 10–2.

10. Change the original circuit to the new circuit shown in Figure 10–3. In the space provided below, draw an equivalent circuit by combining the resistors that are in series. Enter the values of the equivalent resistors on your schematic and in Table 10–3.

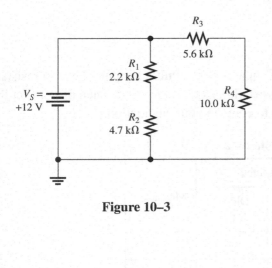

Figure 10–3

Table 10–3

	Computed	Measured
$R_{1,2}$		
$R_{3,4}$		
R_T		
I_T		
$I_{1,2}$		
$I_{3,4}$		
V_1		
V_2		
V_3		
V_4		

11. Compute the resistance of each branch ($R_{1,2}$ and $R_{3,4}$) for the equivalent circuit drawn in step 10. Then compute the total resistance, R_T, of the equivalent circuit. Apply Ohm's law to find the total current I_T. Enter the computed resistance for each branch and the total resistance, R_T, in Table 10–3.

12. Complete the computed values for the circuit by solving for the remaining currents and voltages listed in Table 10–3. Then measure the voltages across each resistor to confirm your computation.

CONCLUSION

EVALUATION AND REVIEW QUESTIONS

1. The voltage divider rule was developed for a series circuit, yet it was applied to the circuit in Figure 10–2.
   (a) Explain.

   (b) Could the voltage divider rule be applied to the circuit in Figure 10–3? Explain your answer.

2. As a check on your solution of the circuit in Figure 10–3, apply Kirchhoff's voltage law to each of two separate paths around the circuit. Show the application of the law.

3. Show the application of Kirchhoff's current law to the junction of R_2 and R_4 of the circuit in Figure 10–3.

4. In the circuit of Figure 10–3, assume you found that I_T was the same as the current in R_3 and R_4.
   (a) What are the possible problems?

   (b) How would you isolate the specific problem using only a voltmeter?

5. The circuit in Figure 10–4 has three equal resistors. If the voltmeter reads $+8.0$ V, find V_S.

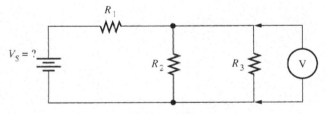

Figure 10–4

FOR FURTHER INVESTIGATION

Figure 10–5 illustrates another series-parallel circuit using the same resistors. Develop a procedure for solving the currents and voltages throughout the circuit. Summarize your procedure in a laboratory report. Confirm your method by computing and measuring the voltages in the circuit.

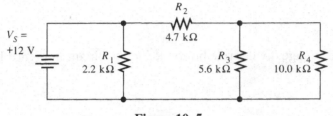

Figure 10–5

MULTISIM TROUBLESHOOTING

This experiment has four Multisim files on the website (www.prenhall.com/floyd). Three of the four files contain a simulated "fault"; one has "no fault". The file with no fault is named EXP10-2-nf. You may want to open this file to compare your results with the computer simulation. Then open each of the files with faults. Use the simulated instruments to investigate the circuit and determine the problem. The following are the filenames for circuits with troubleshooting problems for this experiment.

EXP10-2-f1

    Fault: _____

EXP10-2-f2

    Fault: _____

EXP10-2-f3

    Fault: _____

11 The Superposition Theorem

Name _____
Date _____
Class _____

READING
Text, Section 6–8

OBJECTIVES
After performing this experiment, you will be able to:
1.   Apply the superposition theorem to linear circuits with more than one voltage source.
2.   Construct a circuit with two voltage sources, solve for the currents and voltages throughout the circuit, and verify your computation by measurement.

MATERIALS NEEDED
Resistors:
   One 4.7 kΩ, one 6.8 kΩ, one 10.0 kΩ

SUMMARY OF THEORY
To superimpose something means to lay one thing on top of another. The superposition theorem is a means by which we can solve circuits that have more than one independent voltage source. Each source is taken, one at a time, as if it were the only source in the circuit. All other sources are replaced with their internal resistance. (The internal resistance of a dc power supply or battery can be considered to be zero.) The currents and voltages for the first source are computed. The results are marked on the schematic, and the process is repeated for each source in the circuit. When all sources have been taken, the overall circuit can be solved. The algebraic sum of the superimposed currents and voltages is computed. Currents that are in the same direction are added; those that are in opposing directions are subtracted with the sign of the larger applied to the result. Voltages are treated in a like manner.

The superposition theorem will work for any number of sources *as long as you are consistent in accounting for the direction of currents and the polarity of voltages.* One way to keep the accounting straightforward is to assign a polarity, right or wrong, to each component. Tabulate any current which is in the same direction as the assignment as a positive current and any current which opposes the assigned direction as a negative current. When the final algebraic sum is completed, positive currents are in the assigned direction; negative currents are in the opposite direction of the assignment. In the process of replacing a voltage source with its zero internal resistance, you may completely short out a resistor in the circuit. If this occurs, there will be no current in that resistor for this part of the calculation. The final sum will still have the correct current.

PROCEDURE

1. Obtain the resistors listed in Table 11–1. Measure each resistor and record the measured value in Table 11–1.

2. Construct the circuit shown in Figure 11–1. This circuit has two voltage sources connected to a common reference ground.

Table 11–1

	Listed Value	Measured Value
R_1	4.7 kΩ	
R_2	6.8 kΩ	
R_3	10.0 kΩ	

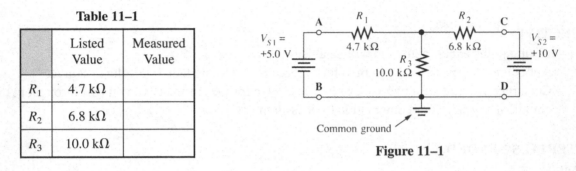

Figure 11–1

3. Remove the 10 V source and place a jumper between the points labeled **C** and **D,** as shown in Figure 11–2. This jumper represents the internal resistance of the 10 V power supply.

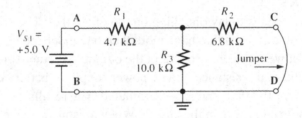

Figure 11–2

4. Compute the total resistance, R_T, seen by the +5.0 V source. Then temporarily remove the +5.0 V source and measure the resistance between points **A** and **B** to confirm your calculation. Record the computed and measured values in Table 11–2.

Table 11–2 Computed and measured resistances.

	Quantity	Computed	Measured
Step 4	R_T (V_{S1} operating alone)		
Step 7	R_T (V_{S2} operating alone)		

5. Use the source voltage, V_{S1}, and the total resistance to compute the total current, I_T, from the +5.0 V source. This current is through R_1, so record it as I_1 in Table 11–3. Use the current divider rule to determine the currents in R_2 and R_3. The current divider rule for I_2 and I_3 is:

$$I_2 = I_T \left(\frac{R_3}{R_2 + R_3} \right) \qquad I_3 = I_T \left(\frac{R_2}{R_2 + R_3} \right)$$

98

Table 11–3 Computed and measured current and voltage.

	Computed Current			Computed Voltage			Measured Voltage		
	I_1	I_2	I_3	V_1	V_2	V_3	V_1	V_2	V_3
Step 5									
Step 6									
Step 8									
Step 9									
Step 10 (totals)									

Record all three currents as *positive* values in Table 11–3. This will be the assigned direction of current. Mark the magnitude and direction of the current in Figure 11–2. Note that the current divider rules shown in this step are only valid for this particular circuit.

6.   Use the currents computed in step 5 and the measured resistances to calculate the expected voltage across each resistor of Figure 11–2. Then connect the +5.0 V power supply and measure the actual voltages present in this circuit. Record the computed and measured voltages in Table 11–3. Since all currents in step 5 were considered *positive*, all voltages in this step are also *positive*.

7.   Remove the +5.0 V source from the circuit and move the jumper from between points **C** and **D** to between points **A** and **B**. Compute the total resistance between points **C** and **D**. Measure the resistance to confirm your calculation. Record the computed and measured resistance in Table 11–2.

8.   Compute the current through each resistor in Figure 11–3. Note that this time the total current is through R_2 and divides between R_1 and R_3. Mark the magnitude and direction of the current on Figure 11–3. *Important:* Record the current as a *positive* current if it is in the same direction as recorded in step 5 and as a *negative* current if it is in the opposite direction as in step 4. Record the computed currents in Table 11–3.

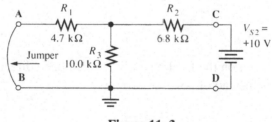

Figure 11–3

9.   Use the currents computed in step 8 and the measured resistances to compute the voltage drops across each resistor. Record the computed voltage drops in Table 11–3. If the current through a resistor was a *positive* current, record the resistor's voltage as a *positive* voltage. If a current was a *negative* current, record the voltage as a *negative* voltage. Then connect the +10 V source as illustrated in Figure 11–3, measure, and record the voltages. The measured voltages should confirm your calculation.

99

10. Compute the algebraic sum of the currents and voltages listed in Table 11–3. Enter the computed sums in Table 11–3. Then replace the jumper between **A** and **B** with the +5.0 V source, as shown in the original circuit in Figure 11–1. Measure the voltage across each resistor in this circuit. The measured voltages should agree with the algebraic sums. Record the measured results in Table 11–3.

CONCLUSION

EVALUATION AND REVIEW QUESTIONS
1. (a) Prove that Kirchhoff's voltage law is valid for the circuit in Figure 11–1. Do this by substituting the measured algebraic sums from Table 11–3 into a loop equation written around the outside loop of the circuit.

   (b) Prove Kirchhoff's current law is valid for the circuit of Figure 11–1 by writing an equation showing the currents entering a junction are equal to the currents leaving the junction. Keep the assigned direction of current from step 5 and use the signed currents computed in step 10.

2. If an algebraic sum in Table 11–3 is negative, what does this indicate?

3. What would be the effect on the final result if you had been directed to record all currents in step 5 as negative currents instead of positive currents?

4.   In your own words, list the steps required to apply the superposition theorem.

5.   Use the superposition theorem to find the current in R_2 in Figure 11–4.

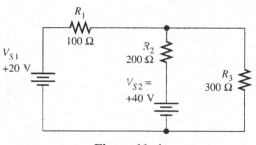

Figure 11–4

FOR FURTHER INVESTIGATION

Compute the power dissipated in each resistor in the circuits shown in Figures 11–1, 11–2, and 11–3. Using the computed results, find out if the superposition theorem is valid for power. Summarize your computations and conclusion.

MULTISIM TROUBLESHOOTING

This experiment has four Multisim files on the website (www.prenhall.com/floyd). Three of the four files contain a simulated "fault"; one has "no fault". The file with no fault is named EXP11-1-nf. You may want to open this file to compare your results with the computer simulation. Then open each of the files with faults. Use the simulated instruments to investigate the circuit and determine the problem. The following are the filenames for circuits with troubleshooting problems for this experiment.

EXP11-1-f1
      Fault:

EXP11-1-f2
      Fault:

EXP11-1-f3
      Fault:

12 Thevenin's Theorem

Name _____
Date _____
Class _____

READING
Text, Section 6–6

OBJECTIVES
After performing this experiment, you will be able to:
1. Change a linear network containing several resistors into an equivalent Thevenin circuit.
2. Prove the equivalency of the network in objective 1 with the Thevenin circuit by comparing the effects of various load resistors.

MATERIALS NEEDED
Resistors:
    One 150 Ω, one 270 Ω, one 470 Ω, one 560 Ω, one 680 Ω, one 820 Ω
One 1 kΩ potentiometer

SUMMARY OF THEORY
In Experiment 10, you solved series-parallel circuits by developing equivalent circuits. Equivalent circuits simplify the task of solving for current and voltage in a network. The concept of equivalent circuits is basic to solving many problems in electronics.

Thevenin's theorem provides a means of reducing a complicated, linear network into an equivalent circuit when there are two terminals of special interest (usually the output). The equivalent Thevenin circuit is composed of a voltage source and a series resistor. (In ac circuits, the resistor may be represented by opposition to ac called *impedance*.) Imagine a complicated network containing multiple voltage sources, current sources, and resistors, such as that shown in Figure 12–1(a). Thevenin's theorem can reduce this to the equivalent circuit shown in Figure 12–1(b). The circuit in Figure 12–1(b) is called a Thevenin circuit. A device connected to the output is a *load* for the Thevenin circuit. The two circuits have identical responses to any load.

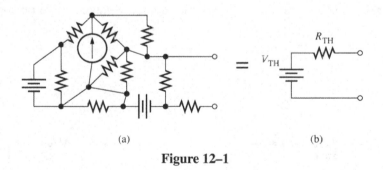

(a)   (b)

Figure 12–1

Two steps are required in order to simplify a circuit to its equivalent Thevenin circuit. The first step is to measure or compute the voltage at the output terminals with any load resistors removed. This open-circuit voltage is the Thevenin voltage. The second step is to compute the resistance seen at the same open terminals if sources are replaced with their internal resistance. For voltage sources, the internal resistance is usually taken as zero, and for current sources, the internal resistance is infinite (open circuit). An example of this process is illustrated in Figure 12–2.

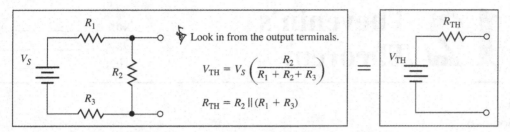

Important: The equations developed in this example are given to illustrate a *procedure* and are valid *only* for the example; they cannot be applied to other circuits, including the circuit in this experiment.

Figure 12–2

PROCEDURE

1. Measure and record the resistance of the 6 resistors listed in Table 12–1. The last three resistors will be used as load resistors and connected, one at a time, to the output terminals.

Table 12–1

Component	Listed Value	Measured Value
R_1	270 Ω	
R_2	560 Ω	
R_3	680 Ω	
R_{L1}	150 Ω	
R_{L2}	470 Ω	
R_{L3}	820 Ω	

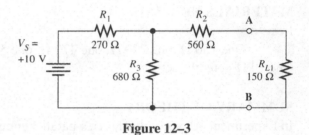

Figure 12–3

2. Construct the circuit shown in Figure 12–3. Points **A** and **B** represent the output terminals. Calculate an equivalent circuit seen by the voltage source. Figure 12–4 illustrates the procedure. Use the equivalent circuit to compute the expected voltage across the load resistor, V_{L1}. Do not use Thevenin's theorem at this time. Show your computation of the load voltage in the space provided. For the first load resistor, R_{L1}, your computed voltage should be approximately 1.19 V.

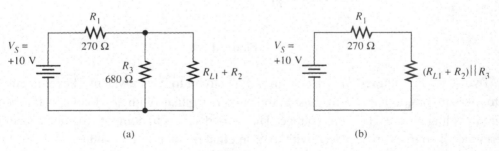

(a)   (b)

Figure 12–4

104

3. Measure the load voltage to verify your calculation. Enter the computed and measured load voltage in Table 12–2.

Table 12–2

	Computed	Measured
V_{L1}		
V_{L2}		
V_{L3}		
V_{TH}		
R_{TH}		

4. Replace R_{L1} with R_{L2}. Using a new equivalent circuit, compute the expected voltage, V_{L2}, across the load resistor. Then measure the actual load voltage. Enter the computed and measured voltage in Table 12–2.

5. Repeat step 4 using R_{L3} for the load resistor.

6. Remove the load resistor from the circuit. Calculate the open circuit voltage at the **A–B** terminals. This open circuit voltage is the *Thevenin voltage* for this circuit. Record the open circuit voltage in Table 12–2 as V_{TH}.

7. Mentally replace the voltage source with a short (zero ohms). Compute the resistance between the **A–B** terminals. This is the computed *Thevenin resistance* for this circuit. Then disconnect the voltage source and replace it with a jumper. Measure the actual Thevenin resistance of the circuit. Record your computed and measured Thevenin resistance in Table 12–2.

8. In the space below, draw the Thevenin equivalent circuit. Show on your drawing the measured Thevenin voltage and resistance.

9. For the circuit you drew in step 8, compute the voltage you expect across each of the three load resistors. Since the circuit is a series circuit, the voltage divider rule will simplify the calculation. Enter the computed voltages in Table 12–3.

Table 12–3

	Computed	Measured
V_{L1}		
V_{L2}		
V_{L3}		
V_{TH}		
R_{TH}		

10. Construct the Thevenin circuit you drew in step 8. Use a 1 kΩ potentiometer to represent the Thevenin resistance. Set it for the resistance shown on your drawing. Set the voltage source for the Thevenin voltage. Place each load resistor, one at a time, on the Thevenin circuit and measure the load voltage. Enter the measured voltages in Table 12–3.

11. Remove the load resistor from the Thevenin circuit. Find the open circuit voltage with no load. Enter this voltage as the computed and measured V_{TH} in Table 12–3. Enter the measured setting of the potentiometer as R_{TH} in Table 12–3.

CONCLUSION

EVALUATION AND REVIEW QUESTIONS

1. Compare the measured voltages in Tables 12–2 and 12–3. What conclusion can you draw about the two circuits?

2. Compute the load current you would expect to measure if the load resistor in Figure 12–3 were replaced with a short. Then repeat the computation for the Thevenin circuit you drew in step 8.

3. What advantage does Thevenin's theorem offer for computing the load voltage across each of the load resistors tested in this experiment?

4. Figure 12–5(a) shows a circuit, and Figure 12–5(b) shows its equivalent Thevenin circuit. Explain why R_1 has no effect on the Thevenin circuit.

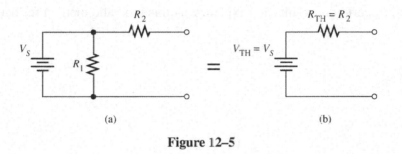

(a)   (b)

Figure 12–5

5. Draw the Thevenin circuit for the circuits shown in Figure 12–6.

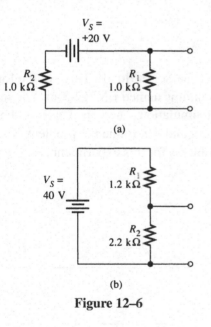

(a)

(b)

Figure 12–6

FOR FURTHER INVESTIGATION

Sometimes it is useful to compute a Thevenin equivalent circuit when it is not possible to measure the Thevenin resistance directly. The Thevenin resistance can still be determined for a source by placing a known load resistor on the output terminals and observing the loaded and unloaded output voltage. A simple method is to use a variable resistor as a load resistor and adjust it until the load voltage has dropped to one-half of the open circuit voltage. The variable load resistor and the internal Thevenin resistance of the source will then be equal. Use this method to measure the Thevenin resistance of your signal generator. Report your results and explain why this is a valid method for determining the Thevenin resistance.

MULTISIM TROUBLESHOOTING

This experiment has four Multisim files on the website (www.prenhall.com/floyd). Three of the four files contain a simulated "fault"; one has "no fault". The file with no fault is named EXP12-3-nf. You may want to open this file to compare your results with the computer simulation. Then open each of the files with faults. Use the simulated instruments to investigate the circuit and determine the problem. The following are the filenames for circuits with troubleshooting problems for this experiment.

EXP12-3-f1
    Fault: _____

EXP12-3-f2
    Fault: _____

EXP12-3-f3
    Fault: _____

13 The Wheatstone Bridge

Name _____
Date _____
Class _____

READING
Text, Sections 6–5 through 6–8

OBJECTIVES
After performing this experiment, you will be able to:
1.   Calculate the equivalent Thevenin circuit for a Wheatstone bridge circuit.
2.   Verify that the Thevenin circuit determined in objective 1 enables you to compute the response to a load for the original circuit.
3.   Balance a Wheatstone bridge and draw the Thevenin circuit for the balanced bridge.

MATERIALS NEEDED
Resistors:
        One 100 Ω, one 150 Ω, one 330 Ω, one 470 Ω
One 1 kΩ potentiometer
For Further Investigation:
    Wheatstone bridge sensitive to 0.1 Ω

SUMMARY OF THEORY
The Wheatstone bridge is a circuit with wide application in measurement systems. It can be used to accurately compare an unknown resistance with known precision resistors and is very sensitive to changes in the unknown resistance. The unknown resistance is frequently a transducer such as a strain gauge, in which very small changes in resistance are related to mechanical stress. The basic Wheatstone bridge is shown in Figure 13–1(a).

Thevenin's theorem is very useful for analysis of the Wheatstone bridge, which is not a simple series-parallel combination circuit. From the perspective of the current in the load resistor, the method shown in the text is the most straightforward analysis technique. The following alternate method is very similar but preserves the ground reference point of the voltage source. This simplifies finding *all* of the currents in the bridge and finding the voltage at point **A** or **B** with respect to ground.

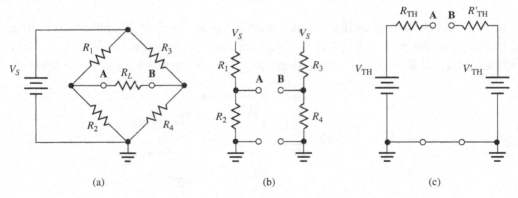

(a)   (b)   (c)

Figure 13–1

Begin by splitting the bridge into two independent voltage dividers as shown in Figure 13–1(b). Thevenin's theorem is applied between point **A** and ground for the left divider and between point **B** and ground for the right divider. V_A is the Thevenin voltage for the left divider, and V_B is the Thevenin voltage for the right divider. To find the Thevenin resistance, the source is replaced with a short, and the resistors on each side are seen to be in parallel. Two Thevenin circuits are then drawn as shown in Figure 13–1(c).

The load resistor can be added to the equivalent circuit as shown in Figure 13–2. Load current can be quickly found by the superposition theorem. The equations for the procedure are given for reference in Figure 13–2.

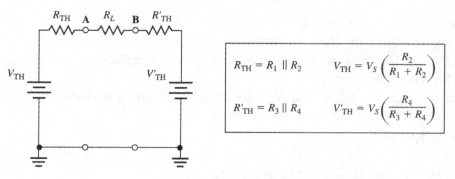

$$R_{TH} = R_1 \| R_2 \qquad V_{TH} = V_S\left(\frac{R_2}{R_1 + R_2}\right)$$

$$R'_{TH} = R_3 \| R_4 \qquad V'_{TH} = V_S\left(\frac{R_4}{R_3 + R_4}\right)$$

Figure 13–2

PROCEDURE

1. Measure and record the resistance of each of the four resistors listed in Table 13–1. R_4 is a 1 kΩ potentiometer. Set it for its maximum resistance and record this value.

Table 13–1

Component	Listed Value	Measured Value
R_1	100 Ω	
R_2	150 Ω	
R_3	330 Ω	
R_L	470 Ω	
R_4	1 kΩ pot.	

2. Construct the Wheatstone bridge circuit shown in Figure 13–3. R_4 should be set to its maximum resistance. Use the voltage divider rule to compute the voltage at point **A** with respect to ground and the voltage at **B** with respect to ground. Enter the computed V_A and V_B in Table 13–2.

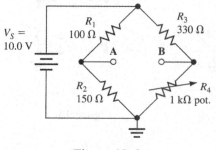

Figure 13–3

110

Table 13–2

	Computed	Measured
V_A		
V_B		
R_{TH}		
R'_{TH}		
V_L		

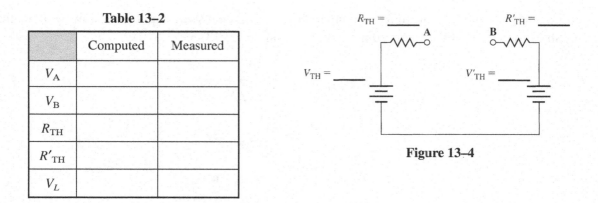

Figure 13–4

3. Measure V_A and V_B. Because these voltages are measured with no load, they are the Thevenin voltages for the bridge using the method illustrated in Figure 13–1. Enter the measured voltages in Table 13–2 and show them on Figure 13–4.

4. Compute the Thevenin resistance on the left side of the bridge in Figure 13–3 by mentally replacing V_S with a short. Notice that this causes R_1 to be in parallel with R_2. Repeat the process for the right side of the bridge. Enter the computed Thevenin resistances, R_{TH} and R'_{TH}, in Table 13–2 and on Figure 13–4. Then replace V_S with a short and measure R_{TH} and R'_{TH}.

5. Draw in the load resistor between the **A** and **B** terminals in the circuit of Figure 13–4. Show the value of the measured resistance of R_L. Use the superposition theorem to compute the expected voltage drop, V_L, across the load resistor. Enter the computed voltage drop in Table 13–2.

6. Place the load resistor across the **A** and **B** terminals of the bridge circuit (Figure 13–3) and measure the load voltage, V_L. If the measured value does not agree with the computed value, recheck your work. Enter the measured V_L in Table 13–2.

7. Monitor the voltage across the load resistor and carefully adjust R_4 until the bridge is balanced. When balance is achieved, remove the load resistor. Measure the voltage from **A** to ground and the voltage from **B** to ground. Since the load resistor has been removed, these measurements represent the Thevenin voltages of the balanced bridge. Enter the measured voltages on Figure 13–5.

8. Replace the voltage source with a short. With the short in place, measure the resistance from point **A** to ground and from point **B** to ground. Enter the measured resistances on Figure 13–5.

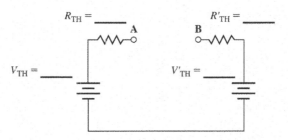

Figure 13–5 Thevenin circuit for balanced bridge.

111

9. Use the superposition theorem to combine the two Thevenin sources into one equivalent circuit. Show the values of the single equivalent circuit on Figure 13–6.

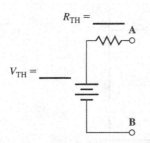

Figure 13–6 Net Thevenin circuit for balanced bridge.

CONCLUSION

EVALUATION AND REVIEW QUESTIONS

1. If you doubled the load resistor in a Wheatstone bridge, the load current would *not* be half as much. Why not?

2. (a) Does a change in the load resistor change the currents in the arms of an *unbalanced* bridge?

    (b) Does a change in the load resistor change the currents in the arms of a *balanced* bridge? Explain.

3.　(a)　What would happen to the load current of an *unbalanced* bridge if all the bridge resistors were doubled in size?

　　(b)　What would happen to the load current of a *balanced* bridge if all the bridge resistors were doubled in size?

4.　(a)　What would happen to the load current of an *unbalanced* bridge if the source voltage were doubled?

　　(b)　What would happen to the load current of a *balanced* bridge if the source voltage were doubled?

5.　Compute the load current for the bridge in Figure 13–7. Show your work.

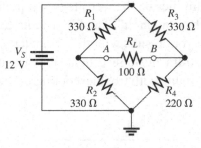

Figure 13–7

FOR FURTHER INVESTIGATION

To do this investigation, you will need a calibrated Wheatstone bridge, capable of making resistance measurements within 0.1 Ω or better. A Wheatstone bridge can determine the location of a short to ground in a multiple conductor cable. The bridge is connected to make a ratio measurement. Simulate a multiple conductor cable with two small diameter wires (#24 gauge or higher) at least 150 ft long. You will need an accurate total resistance of the wire, which you can obtain from the Wheatstone bridge or a sensitive ohmmeter. Place a short to ground at some arbitrary location along the wire. See Figure 13–8 for a diagram. The wire forms two legs of a Wheatstone bridge as illustrated.

Call the total resistance of the wire r (down and back) and the resistance of the wire to the fault a. The bridge is balanced, and the resistance a is determined by the equation shown in Figure 13–8. The fractional distance to the fault is the ratio of a to $(1/2) r$. If you know the total length of the wire, you can find the distance to the fault by setting up a proportion. Investigate this and report on your results.

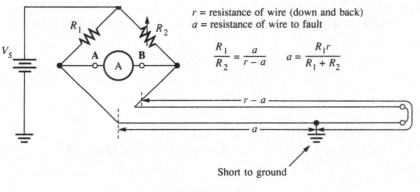

r = resistance of wire (down and back)
a = resistance of wire to fault

$$\frac{R_1}{R_2} = \frac{a}{r-a} \qquad a = \frac{R_1 r}{R_1 + R_2}$$

Short to ground

Figure 13–8

MULTISIM TROUBLESHOOTING

This experiment has four Multisim files on the website (www.prenhall.com/floyd). Three of the four files contain a simulated "fault"; one has "no fault". The file with no fault is named EXP13-7-nf. You may want to open this file to compare your results with the computer simulation. Then open each of the files with faults. Use the simulated instruments to investigate the circuit and determine the problem. The following are the filenames for circuits with troubleshooting problems for this experiment.

EXP13-7-f1
    Fault: _____

EXP13-7-f2
    Fault: _____

EXP13-7-f3
    Fault: _____

Application Assignment 6

Name _____

Date _____

Class _____

REFERENCE

Text, Chapter 6; Application Assignment: Putting Your Knowledge to Work

Step 1   Draw the schematic.

Step 2   Specify how to connect the power supply so that all resistors are in series and pin 2 has the highest voltage.

_____

Steps 3–6   Complete Table AA–6–1. Determine the unloaded output voltages, the loaded output voltages, the percent deviation between the loaded and unloaded voltages, and the load currents.

Table AA–6–1

10 MΩ Load	$V_{OUT\,(2)}$	$V_{OUT\,(3)}$	$V_{OUT\,(4)}$	% Deviation	$I_{LOAD\,(2)}$	$I_{LOAD\,(3)}$	$I_{LOAD\,(4)}$
None							
Pin 2 to ground							
Pin 3 to ground							
Pin 4 to ground							
Pin 2 to ground				2			
Pin 3 to ground				3			
Pin 2 to ground				2			
Pin 4 to ground				4			
Pin 3 to ground				3			
Pin 4 to ground				4			
Pin 2 to ground				2			
Pin 3 to ground				3			
Pin 4 to ground				4			

Specify the minimum value for the fuse: _____

115

Step 7   Troubleshooting:
                  Case 1: _____
                  Case 2: _____
                  Case 3: _____
                  Case 4: _____
                  Case 5: _____
                  Case 6: _____
                  Case 7: _____
                  Case 8: _____

RELATED EXPERIMENT

MATERIALS NEEDED
Resistors:
        One 68 Ω, one 100 Ω, one 560 Ω

DISCUSSION
The application assignment involved determining the effects of a load on a voltage divider. Similar effects occur with a resistive matching network. A circuit is designed to match the resistance of a source and load. A circuit that performs this function is called an *attenuator pad*. The L-section shown in Figure AA–6–1 is a loaded voltage divider designed to match a higher resistance to a lower resistance. The load resistance is taken into account in the design of the divider network. The total resistance looking into the L-pad is very close to 600 Ω, the same as the source resistance.

    Construct the circuit, and connect a 600 Ω source. The source can be a signal generator with an internal 600 Ω resistance set for a 1.0 kHz sine wave or a dc power supply with a series 600 Ω resistor. Set the source voltage to 5.0 V with a source open. Then connect the L-pad and load and observe V_{in} and V_{out}. What happens to V_{in} when the L-pad and load are connected? Compute and measure the attenuation (ratio of V_{out} to V_{in}). Is there a whole-number ratio between the output voltage and the input voltage? Write a short report on your results.

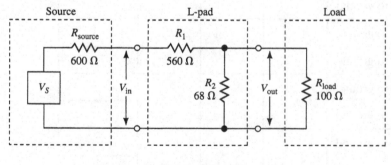

Figure AA–6–1

116

Checkup 6

REFERENCE

Text, Chapter 6; Lab manual, Experiments 10, 11, 12, and 13

1. The term that best describes the analysis of a series-parallel circuit is:
   (a) one current   (b) same voltage   (c) equivalent circuits   (d) multiple sources

2. If two resistors in a series-parallel circuit are connected in series, the voltage across each will be:
   (a) the same   (b) proportional to the resistance
   (c) inversely proportional to the resistance   (d) equal to the source voltage

3. To minimize loading effects on a voltage divider, the load should be:
   (a) much smaller than the divider resistors   (b) equal to the smallest divider resistor
   (c) equal to the largest divider resistor   (d) much larger than the divider resistors

4. When a load resistor is connected to a voltage divider, the current from the source:
   (a) increases   (b) decreases   (c) stays the same

5. Assume a voltmeter has a sensitivity factor of 10,000 Ω/V. On the 10 V scale, the meter will have an internal resistance of:
   (a) 1000 Ω (b) 10,000 Ω (c) 100 kΩ (d) 1.0 MΩ

6. For the circuit shown in Figure C–6–1, the two resistors that are in series are:
   (a) R_1 and R_2 (b) R_2 and R_3 (c) R_2 and R_4 (d) R_3 and R_4

7. For the circuit shown in Figure C–6–1, the equivalent Thevenin voltage is:
   (a) 1.0 V   (b) 3.0 V   (c) 4.0 V   (d) 12 V

8. For the circuit shown in Figure C–6–1, the equivalent Thevenin resistance is:
   (a) 6.67 kΩ (b) 10 kΩ (c) 16.7 kΩ (d) 30 kΩ

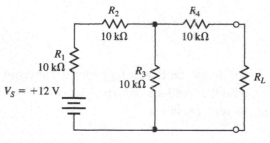

Figure C–6–1

9. To apply the superposition theorem, each source is taken one at a time, as if it were the only source in the circuit. The remaining sources are replaced with:
   (a) their internal resistance   (b) a low resistance
   (c) a high resistance   (d) an open circuit

10. To find the Thevenin voltage of a source, you could measure:
    (a) the voltage across the load   (b) the current in the load
    (c) the load resistance   (d) the open-circuit output voltage

11. Assume a 15 kΩ load resistor is connected to the output terminals of the circuit shown in Figure C–6–1. Compute the voltage and current in the load.

12. Determine the total resistance between the terminals for Figure C–6–2.

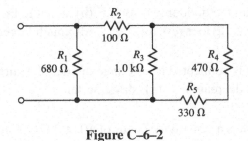

Figure C–6–2

13. In Experiment 10 (Series-Parallel Combination Circuits), you were directed to solve a series-parallel circuit using two methods (Ohm's law and the voltage divider theorem). Why do you think this was requested?

14. In Experiment 11 (The Superposition Theorem), you replaced each source with a jumper wire, one at a time. Is this valid for all sources? Explain.

15. In Experiment 13 (The Wheatstone Bridge), you used Thevenin's theorem to compute parameters for a loaded bridge. Could the methods of Experiment 10 (Series-Parallel Combination Circuits) have been used instead? Why or why not?

14 Magnetic Devices

Name _____
Date _____
Class _____

READING
Text, Sections 7–1 through 7–4

OBJECTIVES
After performing this experiment, you will be able to:
1. Determine the pull-in voltage and release voltage for a relay.
2. Connect relay circuits including a relay latching circuit.
3. Explain the meaning of common relay terminology.

MATERIALS NEEDED
One DPDT relay with a low-voltage dc coil
Two LEDs: one red, one green
One SPST switch
Two 330 Ω resistors

SUMMARY OF THEORY
Magnetism plays an important role in a number of electronic components and devices including inductors, transformers, relays, solenoids, and transducers. Magnetic fields are associated with the movement of electric charges. By forming a coil, the magnetic field lines are concentrated, a fact that is used in most magnetic devices. Wrapping the coil on a core material such as iron, silicon steel, or permalloy provides two additional advantages. First, the magnetic flux is increased because the *permeability* of these materials is much higher than air. Permeability is a measure of how easily magnetic field lines pass through a material. Permeability is not a constant for a material but depends on the amount of flux in the material. The second advantage of using a magnetic core material is that the flux is more concentrated.

A common magnetic device is the *relay*. The relay is an electromagnetic switch with one or more sets of contacts used for controlling large currents or voltages. The switch contacts are controlled by an electromagnet, called the coil. Contacts are specified as either *normally open* (NO) or *normally closed* (NC) when no voltage is applied to the coil. Relays are *energized* by applying the rated voltage to the coil. This causes the contacts to either close or open.

Relays, like mechanical switches, are specified in terms of the number of independent switches (called *poles*) and the number of contacts (called *throws*). Thus, a single-pole double-throw (SPDT) relay has a single switch with two contacts—one normally open and one normally closed. An example of such a relay in a circuit is shown in Figure 14–1. With S_1 open, the motor is off and the light is on. When S_1 is closed, coil CR_1 is energized, causing the NO contacts to close and the NC contacts to open. This applies line voltage to the motor and at the same time removes line voltage from the light. Figure 14–1(a) is drawn in a manner similar to many industrial schematics, sometimes referred to as a ladder schematic. The NC contacts are indicated on this schematic with a diagonal line drawn through them. An alternative way of drawing the schematic is shown in Figure 14–1(b). In this drawing, the relay contacts are drawn as a switch.

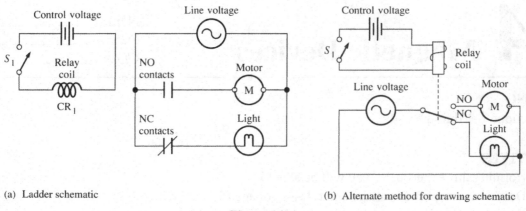

(a) Ladder schematic   (b) Alternate method for drawing schematic

Figure 14–1

Manufacturers specify relays in terms of the ratings for the coil voltage and current, maximum contact current, operating time, and so forth. The specification sheet shows the location of contacts and coil. If these are not available, a technician can determine the electrical wiring of contacts and coil by inspection and ohmmeter tests.

PROCEDURE

1. Obtain a double-pole double-throw (DPDT) relay with a low-voltage dc coil. The terminals should be numbered. Inspect the relay to determine which terminals are connected to the coil and which are connected to the contacts. The connection diagram is frequently drawn on the relay. Check the coil with an ohmmeter. It should indicate the coil resistance. Check contacts with the ohmmeter. NC contacts should read near zero ohms, and NO contacts should read infinite resistance. You may have difficulty determining which contact is movable until the coil is energized. In the space provided, draw a diagram of your relay, showing the coil, all contacts, and terminal numbers. Record the coil resistance on your drawing.

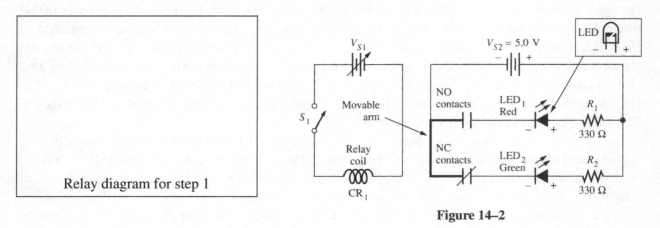

Relay diagram for step 1

Figure 14–2

120

2. Connect the circuit shown in Figure 14–2. In this circuit, only one pole of the relay is used. The movable arm is connected to the negative side of V_{S2}. Note carefully the direction of the light-emitting diodes (LEDs). LEDs are polarized and must be connected in the correct direction. V_{S1} is the control voltage and should be set to the specified coil voltage for the relay. V_{S2} represents a line voltage which is being controlled. For safety, a low voltage is used. Set V_{S2} for 5.0 V. If the circuit is correctly connected, the green LED should be on with S_1 open. Close S_1 and verify that the red LED turns on and the green LED goes off.

3. In this step, you will determine the *pull-in voltage* of the relay. The *pull-in voltage* is the minimum value of coil voltage which will cause the relay to switch. Turn V_{S1} to its lowest setting. With S_1 closed, gradually raise the voltage until the relay trips as indicated with the LEDs. Record the pull-in voltage in Table 14–1.

4. The *release voltage* is the value of the coil voltage at which the contacts return to the unenergized position. Gradually lower the voltage until the relay resets to the unenergized position as indicated by the LEDs. Record the release voltage in Table 14–1.

5. Repeat steps 3 and 4 for two more trials, entering the results of each trial in Table 14–1.

6. Compute the average pull-in voltage and the average release voltage. Enter the averages in Table 14–1.

7. In this step you will learn how to construct a latching relay. Connect the unused NO contacts from the other pole on the relay in parallel with S_1 as illustrated in Figure 14–3. Set V_{S1} for the rated coil voltage. Close and open S_1. Describe your observations.

_____

_____

Table 14–1

		Pull-in Voltage	Release Voltage
Steps 3 and 4	Trial 1		
Step 5	Trial 2		
	Trial 3		
Step 6	Average		

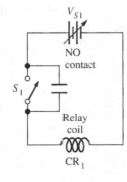
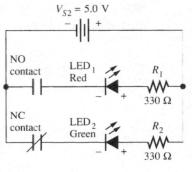

Figure 14–3

121

8.   Remove the NO contact from around S_1. Connect the NC contact in series with S_1 as shown in Figure 14–4. Explain what happens.

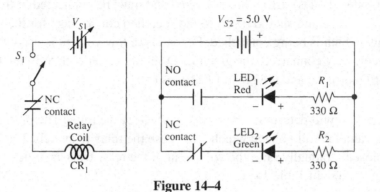

Figure 14–4

CONCLUSION

EVALUATION AND REVIEW QUESTIONS

1.   Using the average pull-in voltage and the measured resistance of your relay coil, compute the average *pull-in current*. The pull-in current is defined as the minimum value of coil current at which the switching function is completed.

2.   Repeat Question 1 for the *release current* using the average of the measured release voltage and the measured resistance of the coil.

3. Hysteresis can be defined as the difference in response due to an increasing or decreasing signal. For a relay, it is the difference between the pull-in and the release voltage. Compute the hysteresis of your relay.

4. (a) Explain the difference between (a) SPDT and (b) DPST.

   (b) Explain the meaning of NO and NC, as it applies to a relay.

5. For the circuit of Figure 14–1, assume that when S is closed, the light stays on and the motor remains off.
   (a) Name two possible faults that could account for this.

   (b) What procedure would you suggest to isolate the fault?

FOR FURTHER INVESTIGATION

A DPDT relay can be used to reverse a voltage—such as causing a dc motor to turn in the opposite direction. Consider the problem of reversing a 5.0 V power supply with a single-pole single-throw switch and a relay, as illustrated in the partial schematic in Figure 14–5. When the switch is closed, the red LED should be ON, but when it is opened, the voltage should reverse, causing the green LED to turn on. Complete the schematic that will accomplish the problem; then build and test your circuit.

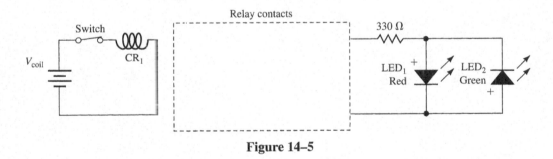

Figure 14–5

Application Assignment 7

Name _____
Date _____
Class _____

REFERENCE

Text, Chapter 7; Application Assignment: Putting Your Knowledge to Work

Step 1 Complete the diagram of the system and provide a wire list.

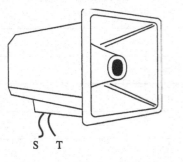

Siren

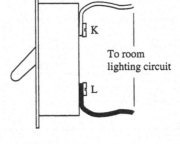

To room
lighting circuit

K

L

Wall switch

M N   O   P   Q   R

Magnetic switches

System ON/OFF
toggle switch

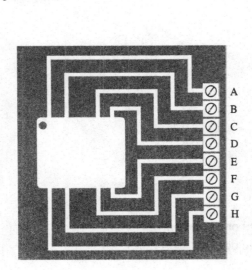

A
B
C
D
E
F
G
H

Relay terminal board

I   J

+   −

Battery

Figure AA–7–1

Wire list. (Two lines are given as an example.)

From	–	To	From	–	To
Relay board–pin A		Relay board–pin F			
Relay board–pin A		Magnetic switch–pin R			

Step 2   Develop a test procedure for the alarm system.

RELATED EXPERIMENT

MATERIALS NEEDED
One CdS photocell (Jameco 120299 or equivalent)
One DPDT relay with a low-voltage dc coil

DISCUSSION
Sometimes security alarm systems, such as the one you figured out in the application assignment, are constructed with a light sensor acting as a switch. Detection is accomplished by breaking a beam of light that is sensed by a photocell. A cadmium sulfide (CdS) photocell is a device that changes its resistance when light strikes it. This change in resistance can be used to energize a relay. Test this idea and devise a circuit in which a CdS cell controls the energizing of a relay. Show the schematic, the measurements you made, and conclusions about your circuit in a short report.

Checkup 7

REFERENCE

Text, Chapter 7; Lab manual, Experiment 14

1. The magnetic field lines that surround a current-carrying wire are:
   (a) parallel to the current
   (b) perpendicular toward the wire
   (c) perpendicular away from the wire
   (d) concentric circles surrounding the wire

2. The magnetic field strength of an electromagnet depends on:
   (a) current in the coil
   (b) type of core material
   (c) number of turns of wire
   (d) all of these

3. The magnetic unit most like resistance in an electrical circuit is:
   (a) reluctance　　(b) magnetic flux　　(c) permeability　　(d) magnetomotive force

4. The magnetic unit most like current in an electrical circuit is:
   (a) reluctance
   (b) magnetic flux
   (c) permeability
   (d) magnetomotive force

5. The tesla is the unit of:
   (a) magnetizing force　　(b) flux　　(c) flux density　　(d) reluctance

6. An electromagnetic device that normally is used to control contact closure in another circuit is a:
   (a) solenoid　　(b) relay　　(c) switch　　(d) transistor

7. The effect that occurs when an increase in field intensity (H) produces little change in flux density (B) is called:
   (a) hysteresis　　(b) saturation　　(c) demagnetization　　(d) permeability

8. The relative permeability of a substance is the ratio of absolute permeability to the permeability of:
   (a) a vacuum　　(b) soft iron　　(c) nickel　　(d) glass

9. The flux density in an iron core depends on the field intensity and the:
   (a) area　　(b) length　　(c) permeability　　(d) retentivity

10. Assume a coil with an mmf of 500 ampere-turns (At) has a flux of 100 μWb. The reluctance is:
    (a) 5×10^6 At/Wb
    (b) 0.2×10^{-6} At/Wb
    (c) 0.5 At/Wb
    (d) 5×10^{-6} At/Wb

11.   Show how to use one set of contacts on a DPDT relay to form a latching relay.

12.   In Experiment 14 (Magnetic Devices), you observed that the release voltage of a relay is less than the pull-in voltage. Explain why this is true.

13.   (a)   Compare the magnetic field strength of a 1000-turn coil that contains 100 mA of current with a 2000-turn coil that contains 50 mA of current.

       (b)   Compare the flux intensity of the two coils, assuming they are both the same length.

14.   Explain how Faraday's law accounts for the voltage from a basic dc generator.

15.   Assume a flux of 500 μWb is distributed evenly across a rectangular area that is 10 cm $\times$ 10 cm.
       (a)   What is the flux density?

       (b)   How much of the flux in (a) will pass through a 1 cm $\times$ 1 cm square?

15 The Oscilloscope

Name _____
Date _____
Class _____

READING
Text, Section 8–9
Oscilloscope Guide, Lab manual pages 7 through 14

OBJECTIVES
After performing this experiment, you will be able to:
1.   Explain the four functional blocks on an oscilloscope and describe the major controls within each block.
2.   Use an oscilloscope to measure ac and dc voltages.

MATERIALS NEEDED
None

SUMMARY OF THEORY
The oscilloscope is an extremely versatile instrument that lets you see a picture of the voltage in a circuit as a function of time. There are two basic types of oscilloscopes—analog oscilloscopes and digital storage oscilloscopes (DSOs). DSOs are rapidly replacing older analog scopes because they offer significant advantages in measurement capabilities including waveform processing, automated measurements, waveform storage, and printing, as well as many other features. Operation of either type is similar; however, most digital scopes tend to have menus and typically provide the user with information on the display and may have automatic setup provisions.

There is not room in this Summary of Theory to describe all of the controls and features of oscilloscopes, so this is by necessity a limited description. You are encouraged to read the Oscilloscope Guide at the beginning of this manual, which describes the controls in some detail and highlights some of the key differences between analog scopes and DSOs. You can obtain further information from the User Manual packaged with your scope and from manufacturers' websites.

Both analog and digital oscilloscopes have a basic set of four functional groups of controls that you need to be completely familiar with, even if you are using a scope with automated measurements. In this experiment, a generic analog scope is described. Keep in mind, that if you are using a DSO, the controls referred to operate in much the same way but you may see some small operating differences.

Although the process for waveform display is very different between an analog oscilloscope and a DSO, the four main functional blocks and primary controls are equivalent. Figure 15–1 shows a basic analog oscilloscope block diagram which illustrates these four main functional blocks. These blocks are broken down further in the Oscilloscope Guide for both types of scope.

Controls for each of the functional blocks are usually grouped together. Frequently, there are color clues to help you identify groups of controls. Look for the controls for each functional group on your oscilloscope. The display controls include INTENSITY, FOCUS, and BEAM FINDER. The vertical controls include input COUPLING, VOLTS/DIV, vertical POSITION, and channel selection (CH1, CH2, DUAL, ALT, CHOP). The triggering controls include MODE, SOURCE, trigger COUPLING, trigger LEVEL, and others. The horizontal controls include the SEC/DIV, MAGNIFIER, and horizontal POSITION controls. Details of these controls are explained in the referenced reading and in the operator's manual for the oscilloscope.

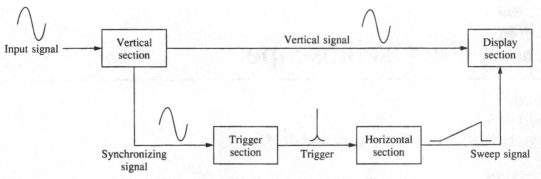

Figure 15–1   Block diagram of an analog oscilloscope

With all the controls to learn, you may experience difficulty obtaining a trace on an analog oscilloscope. If you do not see a trace, start by setting the SEC/DIV control to 0.1 ms/div, select AUTO triggering, select CH1, and press the BEAM FINDER. Keep the BEAM FINDER button depressed and use the vertical and horizontal POSITION controls to center the trace. If you still have trouble, check the INTENSITY control. Note that it's hard to lose the trace on a digital scope, so there is no BEAM FINDER.

Because the oscilloscope can show a voltage-versus-time presentation, it is easy to make ac voltage measurements with a scope. However, care must be taken to equate these measurements with meter readings. Typical digital multimeters show the *rms* (root-mean-square) value of a sinusoidal waveform. This value represents the effective value of an ac waveform when compared to a dc voltage when both produce the same heat (power) in a given load. Usually the *peak-to-peak* value is easiest to read on an oscilloscope. The relationship between the ac waveform as viewed on the oscilloscope and the equivalent rms reading that a DMM will give is illustrated in Figure 15–2.

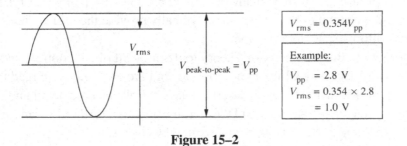

Figure 15–2

Many automated oscilloscopes can measure peak-to-peak or even rms readings of waveforms directly on the screen. They may include horizontal and vertical cursors. Be careful using an automated rms measurement of a sine wave. It may include any dc offset present. If you want to avoid including the dc component, ac couple the signal.

Waveforms that are not sinusoidal cannot be directly compared with an oscilloscope and DMM except for the dc component. The dc level of any waveform can be represented by a horizontal line which splits the waveform into equal areas above and below the line. For a sinusoidal wave, the dc level is always halfway between the maximum and minimum excursions. The dc component can be correctly read by a DMM no matter what the shape of the wave when it is in the DC volts mode.

The amplitude of any periodic waveform can be expressed in one of four ways: the peak-to-peak, the peak, the rms, or the average value. The peak-to-peak value of any waveform is the total magnitude of the change and is *independent* of the zero position. The peak value is the maximum excursion of the wave and is usually referenced to the dc level of the wave. To indicate that a reported value includes a dc offset, you need to state both the maximum and minimum excursions of the waveform.

130

An important part of any oscilloscope measurement is the oscilloscope probe. The type of probe that is generally furnished with an oscilloscope by the manufacturer is called an *attenuator probe* because it attenuates the input by a known factor. The most common attenuator probe is the 10× probe, because it reduces the input signal by a factor of 10. It is a good idea, before making any measurement, to check that the probe is properly compensated, meaning that the frequency response of the probe/scope system is flat. Probes have a small variable capacitor either in the probe tip or a small box that is part of the input connector. This capacitor is adjusted while observing a square wave to ensure that the displayed waveform has vertical sides and square corners. Most oscilloscopes have the square-wave generator built in for the purpose of compensating the probe.

PROCEDURE

1.   Review the front panel controls in each of the major groups. Then turn on the oscilloscope, select CH1, set the SEC/DIV to 0.1 ms/div, select AUTO triggering, and obtain a line across the face of the CRT. Although many of the measurements described in this experiment are automated in newer scopes, it is useful to learn to make these measurements manually.

2.   Turn on your power supply and use the DMM to set the output for 1.0 V. Now we will use the oscilloscope to measure this dc voltage from the power supply. The following steps will guide you:

   (a)   Place the vertical COUPLING (AC-GND-DC) in the GND position. This disconnects the input to the oscilloscope. Use the vertical POSITION control to set the ground reference level on a convenient graticule line near the bottom of the screen.

   (b)   Set the CH1 VOLTS/DIV control to 0.2 V/div. Check that the vernier control is in the CAL position or your measurement will not be accurate. Note that digital scopes do not have a vernier control. For fine adjustments, the VOLTS/DIV control can be changed to a more sensitive setting that remains calibrated.

   (c)   Place the oscilloscope probe on the positive side of the power supply. Place the oscilloscope ground on the power supply common. Move the vertical coupling to the DC position. The line should jump up on the screen by 5 divisions. *Note that 5 divisions times 0.2 V per division is equal to 1.0 V (the supply voltage).* Multiplication of the number of divisions of deflection times volts per division is equal to the voltage measurement.

3.   Set the power supply to each voltage listed in Table 15–1. Measure each voltage using the above steps as a guide. The first line of the table has been completed as an example. To obtain accurate readings with the oscilloscope, it is necessary to select the VOLTS/DIV that gives several divisions of change between the ground reference and the voltage to be measured. The readings on the oscilloscope and meter should agree with each other within approximately 3%.

Table 15–1

Power Supply Setting	VOLTS/DIV Setting	Number of Divisions of Deflection	Oscilloscope (measured voltage)	DMM (measured voltage)
1.0 V	0.2 V/DIV	5.0 DIV	1.0 V	1.0 V
2.5 V				
4.5 V				
8.3 V				

4. Before viewing ac signals, it is a good idea to check the probe compensation for your oscilloscope. To check the probe compensation, set the VOLT/DIV control to 0.1 V/div, the AC-GND-DC coupling control to DC, and the SEC/DIV control to 2 ms/div. Touch the probe tip to the PROBE COMP connector. You should observe a square wave with a flat top and square corners. If necessary, adjust the compensation to achieve a good square wave.

5. Set the function generator for an ac waveform with a frequency of 1.0 kHz. Adjust the amplitude of the function generator for 1.0 V$_{rms}$ as read on your DMM. Set the SEC/DIV control to 0.2 ms/div and the VOLTS/DIV to 0.5 V/div. Connect the scope probe and its ground to the function generator. Adjust the vertical POSITION control and the trigger LEVEL control for a stable display near the center of the screen. You should observe approximately two cycles of an ac waveform with a peak-to-peak amplitude of 2.8 V. This represents 1.0 V$_{rms}$, as shown in Figure 15–3.

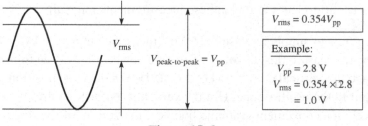

Figure 15–3

6. Use the DMM to set the function generator amplitude to each value listed in Table 15–2. Repeat the ac voltage measurement as outlined in step 5. The first line of the table has been completed as an example. Remember, to obtain accurate readings with the oscilloscope, you should select a VOLTS/DIV setting that gives several divisions of deflection on the screen.

Table 15–2

Signal Generator Amplitude	VOLTS/DIV Setting	Number of Divisions (peak-to-peak)	Oscilloscope Measured (peak-to-peak)	Oscilloscope Measured (rms)
1.0 V$_{rms}$	0.5 V/DIV	5.6 DIV	2.8 V$_{pp}$	1.0 V$_{rms}$
2.2 V$_{rms}$				
3.7 V$_{rms}$				
4.8 V$_{rms}$				

7. Do this step only if you are using an analog oscilloscope. You can observe both the power supply and the function generator at the same time. Select both channels (marked DUAL on some scopes). Each channel can be displayed with its own ground reference point. You will need to leave the trigger SOURCE on channel 2 because the ac waveform is connected to that channel. You can select either ALTernate or CHOP mode to view the waveforms. To really see the effects of this control, slow the function generator to 10 Hz and change the horizontal SEC/DIV control to 20 ms/div. Compare the display using ALTernate and CHOP. At this slow frequency, it is easier to see the waveforms using the CHOP mode; at high frequencies the ALTernate mode is generally preferred.

CONCLUSION

EVALUATION AND REVIEW QUESTIONS

1. (a) Compute the percent difference between the DMM measurement and the oscilloscope measurement for each dc voltage measurement summarized in Table 15–1.

    (b) Which do you think is most accurate? Why?

2. Describe the four major groups of controls on the oscilloscope and the purpose of each group.

3. If you are having difficulty obtaining a stable display, which group of controls should you adjust?

4. (a) If an ac waveform has 3.4 divisions from peak to peak and the VOLTS/DIV control is set to 5.0 V/div, what is the peak-to-peak voltage?

    (b) What is the rms voltage?

5.　　If you wanted to view an ac waveform that was 20.0 V$_{rms}$, what setting of the VOLTS/DIV control would be best?

6.　　Most analog oscilloscopes have a single beam, which is shared with two signals. If you are using an analog oscilloscope, when should you select ALTernate and when should you choose CHOP?

FOR FURTHER INVESTIGATION

Most function generators have a control that allows you to add or subtract a dc offset voltage to the signal. Set up the function generator for a 1.0 kHz sine wave signal, as shown in Figure 15–4. To do this, the AC-GND-DC coupling switch on the oscilloscope should be in the DC position and the offset control should be adjusted on the function generator. When you have the signal displayed on the oscilloscope face, switch the AC-GND-DC coupling switch into the AC position. Explain what this control does. Then measure the signal with your DMM. First measure it in the AC VOLTAGE position; then measure in the DC VOLTAGE position. How does this control differ from the AC-GND-DC coupling switch on the oscilloscope? Summarize your findings.

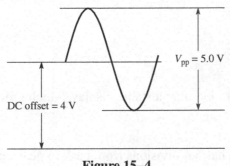

Figure 15–4

16 Sine Wave Measurements

Name _____
Date _____
Class _____

READING
Text, Sections 8–1 through 8–5

OBJECTIVES
After performing this experiment, you will be able to:
1. Measure the period and frequency of a sine wave using an oscilloscope.
2. Measure across ungrounded components using the difference function of an oscilloscope.

MATERIALS NEEDED
Resistors:
    One 2.7 kΩ, one 6.8 kΩ

SUMMARY OF THEORY
Imagine a weight suspended from a spring. If you stretch the spring and then release it, it will bob up and down with a regular motion. The distance from the rest point to the highest (or lowest) point is called the *amplitude* of the motion. As the weight moves up and down, the time for one complete cycle is called a *period,* and the number of cycles it moves in a second is called the *frequency*. This cyclic motion is called *simple harmonic motion*. A graph of simple harmonic motion as a function of time produces a sine wave, the most fundamental waveform in nature. It is also the waveform from an ac generator. Figure 16–1 illustrates these definitions.

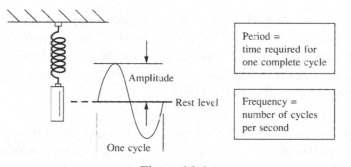

Figure 16–1

Sine waves can also be generated from uniform circular motion. Imagine a circle turning at a constant rate. The *projection* of the endpoint of the radius vector moves with simple harmonic motion. If the end point is plotted along the *x*-axis, the resulting curve is a sine wave, as illustrated in the text. This method is frequently used to show the phase relationship between two sine waves of the same frequency.

The sine wave has another interesting property. Different sine waves can be added together to give new waveforms. In fact, any repeating waveform such as a ramp or square wave can be made up of a group of sine waves. This property is useful in the study of the response of circuits to various waveforms.

The Oscilloscope

As you have seen, there are two basic types of oscilloscopes—analog and digital. In this experiment, you will use an oscilloscope to characterize sine waves. You may want to review the function of the controls on your oscilloscope in the section at the front of this manual entitled Oscilloscope Guide—Analog and Digital Storage Oscilloscopes. Although the method of presenting a waveform is different, the controls such as SEC/DIV are similar in function and should be thoroughly understood. You will make periodic measurements on sine waves in this experiment. Assuming you are not using automated measurements, you need to count the number of divisions for a full cycle and multiply by the SEC/DIV setting to determine the period of the wave. Other measurement techniques will be explained in the Procedure section.

The Function Generator

The basic function generator is used to produce sine, square, and triangle waveforms and may also have a pulse output for testing digital logic circuits. Function generators normally have controls that allow you to select the type of waveform and other controls to adjust the amplitude and dc level. The peak-to-peak voltage is adjusted by the AMPLITUDE control. The dc level is adjusted by a control labeled DC OFFSET; this enables you to add or subtract a dc component to the waveform. These controls are generally not calibrated, so amplitude and dc level settings need to be verified with an oscilloscope or multimeter.

The frequency may be selected with a combination of a range switch and vernier control. The range is selected by a decade frequency switch or pushbuttons that enable you to select the frequency in decade increments (factors of 10) up to about 1 MHz. The vernier control is usually a multiplier dial for adjusting the precise frequency needed.

The output level of a function generator will drop from its open-circuit voltage when it is connected to a circuit. Depending on the conditions, you generally will need to readjust the amplitude level of the generator after it is connected to the circuit. This is because there is effectively an internal generator resistance that will affect the circuit under test. (Review Thevenin's theorem in Experiment 12.)

PROCEDURE

1. Set the signal generator for a 1.0 V_{pp} sine wave at a frequency of 1.25 kHz. Then set the oscilloscope SEC/DIV control to 0.1 ms/div in order to show one complete cycle on the screen. *The expected time for one cycle (the period) is the reciprocal of 1.25 kHz, which is 0.8 ms.* With the SEC/DIV control at 0.1 ms/div, one cycle requires 8.0 divisions across the screen. This is presented as an example in line 1 of Table 16–1.

2. Change the signal generator to each frequency listed in Table 16–1. Complete the table by computing the expected period and then measuring the period with the oscilloscope. Adjust the SEC/DIV control to show between one and two cycles across the screen for each frequency.

Table 16–1

Signal Generator Dial Frequency	Computed Period	Oscilloscope SEC/DIV	Number of Divisions	Measured Period
1.25 kHz	0.8 ms	0.1 ms/div	8.0 div	0.8 ms
1.90 kHz				
24.5 kHz				
83.0 kHz				
600.0 kHz				

136

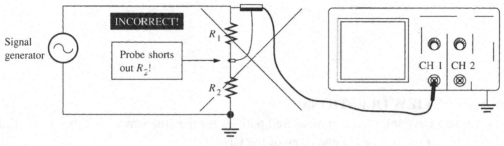

Figure 16–2

3. In this step you will need to use a two-channel oscilloscope with two probes. Frequently, a voltage measurement is needed across an ungrounded component. If the oscilloscope ground is at the same potential as the circuit ground, then the process of connecting the probe will put an undesired ground path in the circuit. Figure 16–2 illustrates this.

The correct way to measure the voltage across the ungrounded component is to use two channels and select the subtract mode—sometimes called the *difference function*—as illustrated in Figure 16–3. The difference function subtracts the voltage measured on channel 2 from the voltage measured on channel 1. It is important that both channels have the same vertical sensitivity—that is, that the VOLTS/DIV setting is the same on both channels and they are both calibrated.

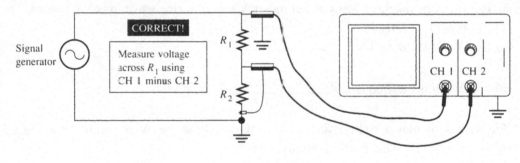

Figure 16–3

Connect the circuit shown in Figure 16–3. Use a 2.7 kΩ resistor for R_1 and a 6.8 kΩ resistor for R_2. Set the signal generator for a 1.0 V$_{pp}$ sine wave at 10 kHz. Channel 1 will show the voltage from the signal generator. Channel 2 will show the voltage across R_2. The difference function (CH1 subtract CH2) will show the voltage across R_1. Some oscilloscopes require that you ADD the channels and INVERT channel 2 in order to measure the difference.[1] Others may have the difference function shown on the Math menu. Complete Table 16–2 for the voltage measurements. Use the voltage divider rule to check that your measured voltages are reasonable.

Table 16–2

	Signal Gen. Voltage	Voltage across R_1	Voltage across R_2
Measured			
Computed	1.0 V$_{pp}$		

---

[1]If you do not have difference channel capability, then temporarily reverse the components to put R_1 at circuit ground. This can be accomplished with a lab breadboard but is not practical in a manufactured circuit.

CONCLUSION

EVALUATION AND REVIEW QUESTIONS

1.   (a)   Compare the computed and measured periods for the sine waves in Table 16–1. Calculate the percent difference for each row of the table.

    (b)   What measurement errors account for the percent differences?

2.   Using the measured voltages in Table 16–2, show that Kirchhoff's voltage law is satisfied.

3.   An oscilloscope display shows one complete cycle of a sine wave in 6.3 divisions. The SEC/DIV control is set to 20 ms/div.
    (a)   What is the period? _____

    (b)   What is the frequency? _____

4.   You wish to display a 10 kHz sine wave on the oscilloscope. What setting of the SEC/DIV control will show one complete cycle in 10 divisions?

    SEC/DIV = _____

5.   Explain how to measure the voltage across an ungrounded component.

FOR FURTHER INVESTIGATION

It is relatively easy to obtain a stable display on the oscilloscope at higher frequencies. It is more difficult to obtain a stable display with slower signals on an analog oscilloscope, especially those with very small amplitude. Set the signal generator on a frequency of 5.0 Hz. Try to obtain a stable display. You will probably have to use NORMAL triggering and carefully adjust the trigger LEVEL control. After you obtain a stable display, try turning the amplitude of the signal generator to its lowest setting. Can you still obtain a stable display?

17 Pulse Measurements

Name _____
Date _____
Class _____

READING
Text, Sections 8–8 and 8–9

OBJECTIVES
After performing this experiment, you will be able to:
1.   Measure rise time, fall time, pulse repetition time, pulse width, and duty cycle for a pulse waveform.
2.   Explain the limitations of instrumentation in making pulse measurements.
3.   Compute the oscilloscope bandwidth necessary to make a rise time measurement with an accuracy of 3%.

MATERIALS NEEDED
One 1000 pF capacitor

SUMMARY OF THEORY
A pulse is a signal that rises from one level to another, remains at the second level for some time, and then returns to the original level. Definitions for pulses are illustrated in Figure 17–1. The time from one pulse to the next is the period, *T*. This is often referred to as the *pulse repetition time*. The reciprocal of period is the *frequency*. The time required for a pulse to rise from 10% to 90% of its maximum level is called the *rise time,* and the time to return from 90% to 10% of the maximum level is called the *fall time.* Pulse width, abbreviated t_w, is measured at the 50% level, as illustrated. The duty cycle is the ratio of the pulse width to the period and is usually expressed as a percentage:

$$\text{Percent duty cycle} = \frac{t_w}{T} \times 100\%$$

Actual pulses differ from the idealized model shown in Figure 17–1(a). They may have *sag, overshoot,* or *undershoot,* as illustrated in Figure 17–1(b). In addition, if cables are mismatched in the system, *ringing* may be observed. Ringing is the appearance of a short oscillatory transient that appears at the top and bottom of a pulse, as illustrated in Figure 17–1(c).

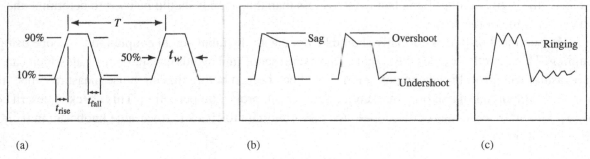

(a)   (b)   (c)

Figure 17–1

All measurements involve some error due to the limitations of the measurement instrument. In this experiment, you will be concerned with rise time measurements. The rise time of the oscilloscope's vertical amplifier (or digitizer's amplifier on a DSO) can distort the measured rise time of a signal. The oscilloscope's rise time is determined by the range of frequencies that can be passed through the vertical amplifier (or digitizing amplifier). This range of frequencies is called the bandwidth, an important specification generally found on the front panel of the scope. Both analog and digital oscilloscopes have internal amplifiers that affect rise time.

If the oscilloscope's internal amplifiers are too slow, rise time distortion may occur, leading to erroneous results. The oscilloscope rise time should be at least four times faster than the signal's rise time if the observed rise time is to have less than 3% error. If the oscilloscope rise time is only twice as fast as the measured rise time, the measurement error rises to over 12%! To find the rise time of an oscilloscope when the bandwidth is known, the following approximate relationship is useful:

$$t_{(r)\text{scope}} = \frac{0.35}{BW}$$

where $t_{(r)\text{scope}}$ is the rise time of the oscilloscope in microseconds and BW is the bandwidth in megahertz. For example, an oscilloscope with a 60 MHz bandwidth has a rise time of approximately 0.006 μs or 6 ns. Measurements of pulses with rise times faster than about 24 ns on this oscilloscope will have measurable error. A correction to the measured value can be applied to obtain the actual rise time of a pulse. The correction formula is

$$t_{(r)\text{true}} = \sqrt{t_{(r)\text{displayed}}^2 + t_{(r)\text{scope}}^2}$$

where $t_{(r)\text{true}}$ is the actual rise time of the pulse, $t_{(r)\text{displayed}}$ is the observed rise time, and $t_{(r)\text{scope}}$ is the rise time of the oscilloscope. This formula can be applied to correct observed rise times by 10% or less.

In addition to the rise time of the amplifier or digitizer, digital scopes have another specification that can affect the usable bandwidth. This specification is the maximum sampling rate. The required sampling rate for a given function depends on a number of variables, but an approximate formula for rise time measurements is

$$\text{Usable bandwith} = \frac{\text{Maximum sampling rate}}{4.6}$$

From this formula, a 1 GHz sampling rate (1 GSa/s) will have a maximum usable bandwidth of 217 MHz. If the digitizer amplifier's bandwidth is less than this, then it should be used to determine the equivalent rise time of the scope.

Measurement of pulses normally should be done with the input signal coupled to the scope using dc coupling. This directly couples the signal to the oscilloscope and avoids causing pulse sag which can cause measurement error. Probe compensation should be checked before making pulse measurements. It is particularly important in rise time measurements to check probe compensation. This check is described in this experiment. For analog oscilloscopes, it is also important to check that variable knobs are in their calibrated position.

PROCEDURE

1. From the manufacturer's specifications, find the bandwidth of the oscilloscope you are using. Normally the bandwidth is specified with a 10× probe connected to the input. You should make oscilloscope measurements with the 10× probe connected to avoid bandwidth reduction. Use the specific bandwidth to compute the rise time of the oscilloscope as explained in the Summary of Theory. This will give you an idea of the limitations of the oscilloscope you are using to make accurate rise time measurements. Enter the bandwidth and rise time of the scope in Table 17–1.

2. Look on your oscilloscope for a probe compensation output. This output provides an internally generated square wave, usually at a frequency of 1.0 kHz. It is a good idea to check this signal when starting with an instrument to be sure that the probe is properly compensated. To compensate the probe, set the VOLTS/DIV control to view the square wave over several divisions of the display. An adjustment screw on the probe is used to obtain a good square wave with a flat top. An improperly compensated oscilloscope will produce inaccurate measurements. If directed by your instructor, adjust the probe compensation.

3. Set the signal generator for a square wave at a frequency of 100 kHz and an amplitude of 4.0 V. A square wave cannot be measured accurately with your meter—you will need to measure the voltage with an oscilloscope. Check the zero volt level on the oscilloscope and adjust the generator to go from zero volts to 4.0 V. Most signal generators have a separate control to adjust the dc level of the signal.

4. Measure the parameters listed in Table 17–2 for the square wave from the signal generator. Be sure the oscilloscope's SEC/DIV is in its calibrated position. If your oscilloscope has percent markers etched on the front gradicule, you may want to *uncalibrate* the VOLTS/DIV when making rise and fall time measurements. Use the vertical POSITION control and VOLTS/DIV vernier to position the waveform between the 0% and 100% markers on the oscilloscope display. Then measure the time between the 10% and 90% markers.[1]

5. To obtain practice measuring rise time, place a 1000 pF capacitor across the generator output. Measure the new rise and fall times. Record your results in Table 17–3.

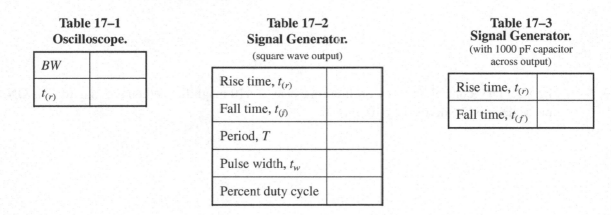

Table 17–1
Oscilloscope.

BW	
$t_{(r)}$	

Table 17–2
Signal Generator.
(square wave output)

Rise time, $t_{(r)}$	
Fall time, $t_{(f)}$	
Period, T	
Pulse width, t_w	
Percent duty cycle	

Table 17–3
Signal Generator.
(with 1000 pF capacitor across output)

Rise time, $t_{(r)}$	
Fall time, $t_{(f)}$	

---

[1]If your oscilloscope has cursor measurements, the rise time can be read directly when the cursors are positioned on the 10% and 90% levels.

6.   If you have a separate pulse output from your signal generator, measure the pulse characteristics listed in Table 17–4. To obtain good results with fast signals, the generator should be terminated in its characteristic impedance (typically 50 Ω). You will need to use the fastest sweep time available on your oscilloscope. Record your results in Table 17–4.

Table 17–4
Signal generator.

(pulse output)

Rise time, $t_{(r)}$	
Fall time, $t_{(f)}$	
Period, T	
Pulse width, t_w	
Percent duty cycle	

CONCLUSION

EVALUATION AND REVIEW QUESTIONS

1.   Were any of the measurements limited by the bandwidth of the oscilloscope? If so, which ones?

2.   If you need to measure a pulse with a predicted rise time of 10 ns, what bandwidth should the oscilloscope have to measure the time within 3%?

3.   The SEC/DIV control on many oscilloscopes has a ×10 magnifier. When the magnifier is ON, the time scale must be divided by 10. Explain.

4.   An oscilloscope presentation has the SEC/DIV control set to 2.0 ms/div and the ×10 magnifier is OFF. Determine the rise time of the pulse shown in Figure 17–2.

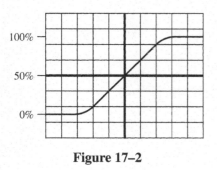

Figure 17–2

5.   Repeat Question 4 for the ×10 magnifier ON.

FOR FURTHER INVESTIGATION

In many applications, it is important to measure time differences. One important technique for doing this with analog oscilloscopes is to use *delayed sweep* measurements. If your scope is equipped with delayed sweep, you can trigger from a signal and view a magnified portion of the signal at a later time. With dual time base oscilloscopes, delayed sweep offers increased timing accuracy. If you have a *calibrated* DELAY TIME POSITION dial, you can make differential delay time measurements between two different signals. Most delayed sweep oscilloscopes will have a HORIZONTAL MODE switch which allows you to view either the A sweep, the B sweep, or A intensified by B. The sweep speeds for A and B can be separately controlled, often by concentric rings on the SEC/DIV control. Consult the operator's manual for your oscilloscope to determine the exact procedure.[2] Then practice by measuring the rise time of the pulse generator using delayed sweep. Summarize your procedure and results.

---

[2]An excellent source of information can be found at http://www.tek.com/. Search for *XYZs of Oscilloscopes,* and several related pdf files will be found.

Application Assignment 8

Name _____
Date _____
Class _____

REFERENCE
Text, Chapter 8; Application Assignment: Putting Your Knowledge to Work

Step 1   Review the operation and controls of the function generator.

Step 2   Measure the sinusoidal output of the function generator.
From Figure 8–68(a)
    minimum amplitude: peak: _____ rms: _____
    minimum frequency _____

From Figure 8–68(b)
    maximum amplitude: peak: _____ rms: _____
    maximum frequency _____

Step 3   Measure the DC offset of the function generator.
From Figure 8–69(a)
    maximum positive dc offset: _____

From Figure 8–69(b)
    maximum negative dc offset: _____

Step 4   Measure the triangular output of the function generator.
From Figure 8–70(a)
    minimum amplitude: _____   minimum frequency _____

From Figure 8–70(b)
    maximum amplitude: _____   maximum frequency _____

Step 5   Measure the pulse output of the function generator.
From Figure 8–71(a)
    minimum amplitude: _____   minimum frequency _____
    duty cycle: _____

From Figure 8–71(b)
    maximum amplitude: _____   maximum frequency _____
    duty cycle: _____

RELATED EXPERIMENT

This application requires you to set up the oscilloscope for optimum settings to measure the period and amplitude of different waveforms. When you measure the period of a signal, choose the lowest SEC/DIV setting that shows at least one full cycle on the display. When measuring amplitude, use the lowest VOLT/DIV setting that shows the entire vertical portion of the waveform. Table AA–8–1 lists waveforms to measure. Before making the measurement, consider the best settings of the controls, and enter the settings in the predicted columns. Set up each signal, measure it, and enter the measured values in the table. Then sketch each waveform on the plots shown, showing the oscilloscope display.

EXPERIMENTAL RESULTS

Table AA–8–1

Function Generator Waveform	Required Amplitude	Required Frequency	VOLTS/DIV Setting (predicted)	SEC/DIV Setting (predicted)	Measured Values of Signal:		
					Horizontal Divisions	Vertical Divisions	Plot Number
Sine wave	$1.0\,V_{rms}$	30 Hz					See Plot AA–8–1
Sine wave	$5.0\,V_{pp}$	30 kHz					See Plot AA–8–2
Pulse	4.0 V	2.5 kHz					See Plot AA–8–3
Pulse	0.5 V	75 kHz					See Plot AA–8–4
Sawtooth	$2.0\,V_{pp}$	400 Hz					See Plot AA–8–5
Sawtooth	$9.0\,V_{pp}$	10 kHz					See Plot AA–8–6

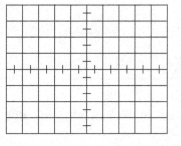

Plot AA–8–1

Plot AA–8–2

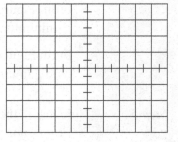

Plot AA–8–3

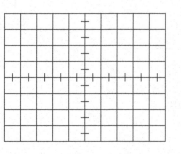

Plot AA–8–4

Plot AA–8–5

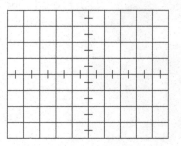

Plot AA–8–6

146

Checkup 8

Name _____

Date _____

Class _____

REFERENCE

Text, Chapter 8; Lab manual, Experiments 15, 16, and 17

1. A sine wave has a peak-to-peak voltage of 25 V. The rms voltage is:
    (a) 8.83 V   (b) 12.5 V   (c) 17.7 V   (d) 35.4 V

2. The number of radians in one-fourth cycle is:
    (a) 57.3   (b) $\pi/2$ (c) π (d) 2π

3. Assume a sine wave has 100 complete cycles in 10 s. The period is:
    (a) 0.1 s   (b) 1 s   (c) 10 s   (d) 100 s

4. Assume a series resistive circuit contains three equal resistors. The source voltage is a sinusoidal waveform of 30 V_{pp}. What is the rms voltage drop across each resistor?
    (a) 3.54 V   (b) 5.0 V   (c) 10 V   (d) 21.2 V

5. Pulse width is normally measured at the:
    (a) 10% level   (b) 50% level   (c) 90% level   (d) baseline

6. A waveform characterized by positive and negative ramps of equal slope is called a:
    (a) triangle   (b) sawtooth   (c) sweep   (d) step

7. A repetitive pulse train has a pulse width of 2.5 μs and a frequency of 100 kHz. The duty cycle is:
    (a) 2.5%   (b) 10%   (c) 25%   (d) 40%

8. The oscilloscope section that determines when it begins to trace a waveform is:
    (a) horizontal   (b) vertical   (c) trigger   (d) display

9. The oscilloscope control that changes the time base is labeled:
    (a) SLOPE   (b) HOLDOFF   (c) VOLTS/DIV   (d) SEC/DIV

10. Measurement of pulses should be done with the input signal coupled to the oscilloscope using
    (a) AC coupling   (b) DC coupling   (c) either ac or dc coupling

11. A standard utility voltage is 120 V at a frequency of 60 Hz.
    (a) What is the peak-to-peak voltage?

    (b) What is the period?

12. How many cycles of a 40 MHz sine wave occur in 0.2 ms?

13. A sinusoidal waveform is represented by the equation $v = 40 \sin(\theta - 35°)$.
    (a) What is the peak voltage?

    (b) What is the phase shift?

14. Figure C–8–1 illustrates an oscilloscope display showing the time relationship between two sine waves. Assume the VOLTS/DIV control is set to 1.0 V/div. Draw a phasor diagram showing the relationship between the two waves.

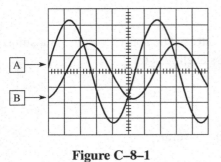

Figure C–8–1

15. A digital oscilloscope has a 1.5 GSa/s sampling rate. What is the usable bandwidth based on the sampling rate?

16. An analog oscilloscope with a bandwidth of 60 MHz is used to measure a pulse with a rise time of 8 ns.
    (a) What is the equivalent rise time of the scope?

    (b) What is the approximate displayed rise time?

17. Why should pulse measurements normally be made with a ×10 probe?

18 Capacitors

READING
Text, Sections 9–1 through 9–5

OBJECTIVES
After performing this experiment, you will be able to:
1.   Compare total capacitance, charge, and voltage drop for capacitors connected in series and in parallel.
2.   Test capacitors with an ohmmeter and a voltmeter.
3.   Determine the value of small capacitors from coded markings.

MATERIALS NEEDED:
Two LEDs
Resistors:
        Two 1.0 kΩ
Capacitors:
        One of each : 100 μF, 47 μF, 1.0 μF, 0.1 μF, 0.01 μF (35 WV or greater)

SUMMARY OF THEORY
A capacitor is formed whenever two conductors are separated by an insulating material. When a voltage exists between the conductors, there will be an electric charge between the conductors. The ability to store an electric charge is a fundamental property of capacitors and affects both dc and ac circuits. Capacitors are made with large flat conductors called *plates*. The plates are separated with an insulating material called a *dielectric*. The ability to store charge increases with larger plate size and closer separation.

When a voltage is connected across a capacitor, charge will flow in the external circuit until the voltage across the capacitor is equal to the applied voltage. The charge that flows is proportional to the size of the capacitor and the applied voltage. This is a fundamental concept for capacitors and is given by the equation

$$Q = CV$$

where Q is the charge in coulombs, C is the capacitance in farads, and V is the applied voltage. An analogous situation is that of putting compressed air into a bottle. The quantity of air is directly proportional to the capacity of the bottle and the applied pressure. (In this analogy, pressure is like voltage, the capacity of the bottle is like capacitance, and the amount of air is like charge.)

Recall that current is defined as charge per time. That is,

$$I = \frac{Q}{t}$$

where I is the current in amperes, Q is the charge in coulombs, and t is the time in seconds. This equation can be rearranged as

$$Q = It$$

If we connect two capacitors in series with a voltage source, the same charging current is through both capacitors. Since this current is for the same amount of time, the total charge, Q_T, must be the same as the charge on each capacitor. That is,

$$Q_T = Q_1 = Q_2$$

Charging capacitors in series causes the same charge to be across each capacitor; however, as shown in the text, the total capacitance *decreases*. In a series circuit, the total capacitance is given by the formula:

$$\frac{1}{C_T} = \frac{1}{C_1} + \frac{1}{C_2} + \dots + \frac{1}{C_n}$$

Now consider capacitors in parallel. In a parallel circuit, the total current is equal to the sum of the currents in each branch as stated by Kirchhoff's current law. If this current is for the same amount of time, the total charge leaving the voltage source will equal the sum of the charges which flow in each branch. That is,

$$Q_T = Q_1 + Q_2 + \dots + Q_n$$

Capacitors connected in parallel will raise the total capacitance because more charge is stored at the same voltage. The equation for the total capacitance of parallel capacitors is:

$$C_T = C_1 + C_2 + \dots + C_n$$

There are two quick tests you can make to check capacitors. The first is an ohmmeter test, useful for capacitors larger than 0.01 μF. This test is best done with an analog ohmmeter rather than a digital meter. The test will sometimes indicate a faulty capacitor is good; however, you can be sure that if a capacitor fails the test, it is bad. The test is done as follows:

(a)   Remove one end of the capacitor from the circuit and discharge it by placing a short across its terminals.

(b)   Set the ohmmeter on a high-resistance scale and place the negative lead from an ohmmeter on the negative terminal of the capacitor. You must connect the ohmmeter with the proper polarity. *Do not assume the common lead from the ohmmeter is the negative side!*

(c)   Touch the other lead of the ohmmeter onto the remaining terminal of the capacitor. The meter should indicate very low resistance and then gradually increase resistance. If you put the meter in a higher range, the ohmmeter charges the capacitor slower and the capacitance "kick" will be emphasized. For small capacitors (under 0.01 μF), this charge may not be seen. Large electrolytic capacitors require more time to charge, so use a lower range on your ohmmeter. Capacitors should never remain near zero resistance, as this indicates a short. An immediate high resistance reading indicates an open for larger capacitors.

A capacitor that passes the ohmmeter test may still fail when voltage is applied. A voltmeter can be used to check a capacitor with voltage applied. The voltmeter is connected in *series* with the capacitor.

When voltage is first applied, the capacitor charges. As it charges, voltage will appear across it, and the voltmeter indication should be a very small voltage. Large electrolytic capacitors may have leakage current that makes them appear bad, especially with a very high impedance voltmeter. As in the case of the ohmmeter test, small capacitors may charge so quickly they appear bad. In these cases, use the test as a relative test, comparing the reading with a similar capacitor that you know is good. Ohmmeter and voltmeter tests are never considered comprehensive tests but are indicative that a capacitor is capable of being charged.

Capacitor Identification

There are many types of capacitors available with a wide variety of specifications for size, voltage rating, frequency range, temperature stability, leakage current, and so forth. For general-purpose applications, small capacitors are constructed with paper, ceramic, or other insulation material and are not polarized. Three common methods for showing the value of a small capacitor are shown in Figure 18–1. In Figure 18–1(a), a coded number is stamped on the capacitor that is read in pF. The first two digits represent the first two digits, the third number is a multiplier. For example, the number 473 is a 47000 pF capacitor. Figure 18–1(b) shows the actual value stamped on the capacitor in μF. In the example shown, .047 μF is the same as 47000 pF. In Figure 18–1(c), a ceramic color-coded capacitor is shown that is read in pF. Generally, when 5 colors are shown, the first is a temperature coefficient (in ppm/°C with special meanings to each color). The second, third, and fourth colors are read as digit 1, digit 2, and a multiplier. The last color is the tolerance. Thus a 47000 pF capacitor will have a color representing the temperature coefficient followed by yellow, violet, and orange bands representing the value. Unlike resistors, the tolerance band is generally green for 5% and white for 10%. More information on capacitor color codes is given in the text in Appendix B.

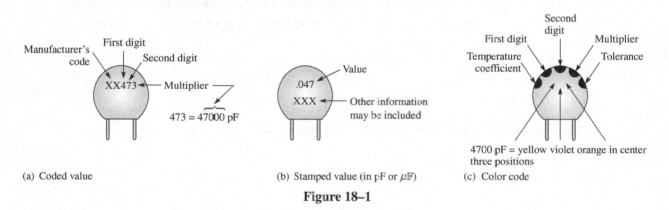

(a) Coded value   (b) Stamped value (in pF or μF)   (c) Color code

Figure 18–1

Larger electrolytic capacitors will generally have their value printed in uncoded form on the capacitor and a mark indicating either the positive or negative lead. They also have a maximum working voltage printed on them which must not be exceeded. Electrolytic capacitors are always polarized, and it is very important to place them into a circuit in the correct direction based on the polarity shown on the capacitor. They can overheat and explode if placed in the circuit backwards.

PROCEDURE

1. Obtain five capacitors as listed in Table 18–1. Check each capacitor using the ohmmeter test described in the Summary of Theory. Record the results of the test in Table 18–1.

2. Test each capacitor using the voltmeter test as illustrated in Figure 18–2. Large electrolytic capacitors or very small capacitors may appear to fail this test, as mentioned in the Summary of Theory. Check the voltage rating on the capacitor to be sure it is not exceeded. The working voltage is the maximum voltage that can safely be applied to the capacitor. Record your results in Table 18–1.

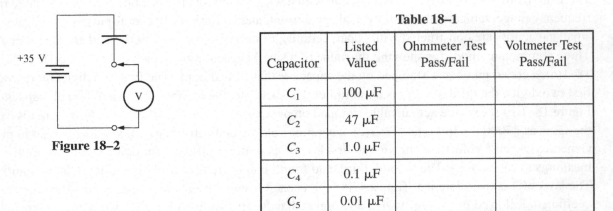

Figure 18–2

Table 18–1

Capacitor	Listed Value	Ohmmeter Test Pass/Fail	Voltmeter Test Pass/Fail
C_1	100 μF		
C_2	47 μF		
C_3	1.0 μF		
C_4	0.1 μF		
C_5	0.01 μF		

3. Connect the circuit shown in Figure 18–3. The switches can be made from jumper wires. Leave both switches open. The light-emitting diodes (LEDs) and the capacitor are both polarized components—they must be connected in the correct direction in order to work properly.

4. Close S_1 and observe the LEDs. Describe your observation.

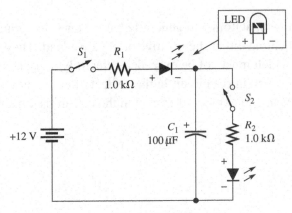

Figure 18–3

152

5. Open S_1 and close S_2. What happens?

6. Now connect C_2 in series with C_1. Open S_2. Make certain the capacitors are fully discharged by shorting them with a piece of wire; then close S_1. Measure the voltage across each capacitor. Do this quickly to prevent the meter from causing the capacitors to discharge. Record the voltages and describe your observations.

   $V_1 = $ _____ $V_2 = $ _____

   Observations:

7. Using the measured voltage, compute the charge on each capacitor.

   $Q_1 = $ _____ $Q_2 = $ _____

   Then open S_1 and close S_2. Observe the result.

8. Change the capacitors from series to parallel. Ensure that the capacitors are fully discharged. Open S_2 and close S_1. Measure the voltage (quickly) across the capacitors. Record the voltages and describe your observations.

   $V_1 = $ _____ $V_2 = $ _____

   Observations:

9. Using the measured voltage, calculate the charge across each capacitor.

   $Q_1 = $ _____ $Q_2 = $ _____

10.　　Replace the 12 V dc source with a signal generator. Close both S_1 and S_2. Set the signal generator to a square wave and set the amplitude to 12 V_{pp}. Set the frequency to 10 Hz. Notice the difference in the LED pulses. This demonstrates one of the principal applications of large capacitors—that of filtering. Explain your observations.

CONCLUSION

EVALUATION AND REVIEW QUESTIONS

1.　　Why did the LEDs flash for a shorter time in step 6 than in steps 4 and 5?

2.　　What would happen if you added more series capacitance in step 6?

3.　　(a)　　What is the total capacitance when a 1.0 μF capacitor is connected in parallel with a 2.0 μF capacitor?

　　　(b)　　If the capacitors are connected in series, what is the total capacitance?

　　　(c)　　In the series connection, which capacitor has the greater voltage across it?

4. Determine the value in pF and µF for each small capacitor with the coded numbers as shown:

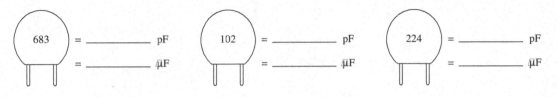

683) = _____ pF
= _____ µF

102) = _____ pF
= _____ µF

224) = _____ pF
= _____ µF

5. Write the coded number that should appear on each capacitor for the values shown:

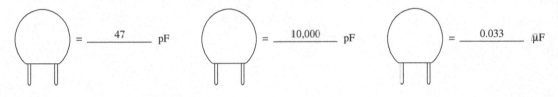

= ____47____ pF

= ____10,000____ pF

= ____0.033____ µF

FOR FURTHER INVESTIGATION

Use the oscilloscope to measure the waveforms across the capacitors and the LEDs in step 10. Try speeding up the signal generator and observe the waveforms. Use the two-channel difference measurement explained in Experiment 16 to see the waveform across the ungrounded LED. Draw and label the waveforms.

Capacitor waveform:

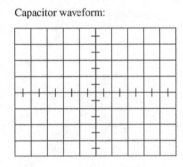

LED waveform:

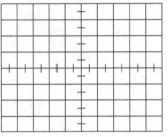

Plot 18–1   **Plot 18–2**

19 Capacitive Reactance

READING

Text, Sections 9–6 and 9–7

OBJECTIVES

After performing this experiment, you will be able to:
1.   Measure the capacitive reactance of a capacitor at a specified frequency.
2.   Compare the reactance of capacitors connected in series and parallel.

MATERIALS NEEDED

One 1.0 kΩ resistor
Two 0.1 µF capacitors
For Further Investigation:
    Two 100 µF capacitors, two LEDs, one 100 kΩ resistor

SUMMARY OF THEORY

If a resistor is connected across a sine wave generator, the current is *in phase* with the applied voltage. If, instead of a resistor, we connect a capacitor across the generator, the current is not in phase with the voltage. This is illustrated in Figure 19–1. Note that the current and voltage have exactly the same frequency, but the current is *leading* the voltage by 1/4 cycle.

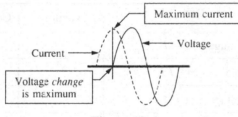

Figure 19–1

Current in the capacitor is directly proportional to the capacitance and the rate of change of voltage. The largest current is when the voltage *change* is a maximum. If the capacitance is increased or the frequency is increased, there is more current. This is why a capacitor is sometimes thought of as a high-frequency short.

Reactance is the opposition to ac current and is measured in ohms, like resistance. Capacitive reactance is written with the symbol X_C. It can be defined as:

$$X_C = \frac{1}{2\pi f C}$$

where f is the generator frequency in hertz and C is the capacitance in farads.

Ohm's law can be generalized to ac circuits. For a capacitor, we can find the voltage across the capacitor using the current and the capacitive reactance. Ohm's law for the voltage across a capacitor is

$$V_C = I X_C$$

PROCEDURE

1. Obtain two capacitors with the values shown in Table 19–1. If you have a capacitance bridge available, measure their capacitance and record in Table 19–1; otherwise, record the listed value of the capacitors. Measure and record the value of resistor R_1.

2. Set up the circuit shown in Figure 19–2. Set the generator for a 1.0 kHz sine wave with a 1.0 V rms output. Measure the rms voltage with your DMM while it is connected to the circuit.[1] Check the frequency and voltage with the oscilloscope. Note: $1.0\,V_{rms} = 2.828\,V_{pp}$.

Table 19–1

Component	Listed Value	Measured Value
C_1	0.1 μF	
C_2	0.1 μF	
R_1	1.0 kΩ	

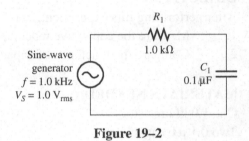

Figure 19–2

3. The circuit is a series circuit, so the current in the resistor is the identical current seen by the capacitor. You can find this current easily by applying Ohm's law to the resistor. Measure the voltage across the resistor, V_R, using the DMM. Record the measured voltage in Table 19–2 in the column labeled Capacitor C_1. Compute the current in the circuit by dividing the measured voltage by the resistance of R_1 and enter in Table 19–2.

Table 19–2

	Capacitor C_1	Capacitor C_2
Voltage across R_1, V_R		
Total current, I		
Voltage across C, V_C		
Capacitive reactance, X_C		
Computed capacitance, C		

---

[1]DMMs have a relatively low bandwidth, although most can measure 1.0 kHz. Verify that the DMM you are using has at least a 1.0 kHz bandwidth; if it does not, use the oscilloscope for all voltage measurements.

4. Measure the rms voltage across the capacitor, V_C. Record this voltage in Table 19–2. Then use this voltage to compute the capacitive reactance using Ohm's law:

$$X_C = \frac{V_C}{I}$$

Enter this value as the capacitive reactance in Table 19–2.

5. Using the capacitive reactance found in step 4, compute the capacitance using the equation

$$C = \frac{1}{2\pi f X_C}$$

Enter the computed capacitance in Table 19–2. This value should agree with the value marked on the capacitor and measured in step 1 within experimental tolerances.

6. Repeat steps 3, 4, and 5 using capacitor C_2. Enter the data in Table 19–2 in the column labeled Capacitor C_2.

7. Now connect C_1 in series with C_2. The equivalent capacitive reactance and capacitance can be found for the series connection by measuring across both capacitors as if they were one capacitor. Enter the data in Table 19–3 in the column labeled Series Capacitors. The following steps will guide you:
   (a) Check that the generator is set to 1.0 V rms. Find the current in the circuit by measuring the voltage across the resistor as before and dividing by the resistance. Enter the measured voltage and the current you found in Table 19–3.
   (b) Measure the voltage across *both* capacitors. Enter this voltage in Table 19–3.
   (c) Use Ohm's law to find the capacitive reactance of both capacitors. Use the voltage measured in step (b) and the current measured in step (a).
   (d) Compute the total capacitance by using the equation

$$C_T = \frac{1}{2\pi f X_{CT}}$$

8. Connect the capacitors in parallel and repeat step 7. Assume the parallel capacitors are one equivalent capacitor for the measurements. Enter the data in Table 19–3 in the column labeled Parallel Capacitors.

Table 19–3

Step		Series Capacitors	Parallel Capacitors
(a)	Voltage across R_1, V_R		
	Total current, I		
(b)	Voltage across capacitors, V_C		
(c)	Capacitive reactance, X_{CT}		
(d)	Computed capacitance, C_T		

CONCLUSION

EVALUATION AND REVIEW QUESTIONS

1.   Compare the capacitive reactance of the series capacitors with the capacitive reactance of the parallel capacitors. Use your data in Table 19–3.

2.   Compare the total capacitance of the series capacitors with the total capacitance of the parallel capacitors.

3.   If someone had mistakenly used too small a capacitor in a circuit, what would happen to the capacitive reactance?

4.   How could you apply the method used in this experiment to find the value of the unknown capacitor?

5.   Compute the capacitive reactance for an 800 pF capacitor at a frequency of 250 kHz.

FOR FURTHER INVESTIGATION

A voltage multiplier is a circuit that uses diodes and capacitors to increase the peak value of a sine wave. Voltage multipliers can produce high voltages without requiring a high-voltage transformer. The circuit illustrated in Figure 19–3 is a full-wave voltage doubler. The circuit is drawn as a bridge with diodes in two arms and capacitors in two arms. The diodes allow current in only one direction, charging the capacitors to near the peak voltage of the sine wave. Generally, voltage doublers are used with 60 Hz power line frequencies and with ordinary diodes, but in order to clarify the operation of this circuit, you can use the LEDs that were used in this experiment. (Note that this causes the output voltage to be reduced slightly.) Connect the circuit, setting the function generator to 20 V_{pp} sine wave at a frequency of 1.0 Hz. (If you cannot obtain a 20 V_{pp} signal, use the largest signal you can obtain from your generator.) Observe the operation of the circuit, then try speeding up the generator. Look at the waveform across the load resistor with your oscilloscope using the two-channel difference method. What is the dc voltage across the load resistor? What happens to the output as the generator is speeded up? Try a smaller load resistor. Can you explain your observations?

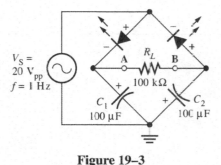

Figure 19–3

MULTISIM TROUBLESHOOTING

This experiment has four Multisim files on the website (www.prenhall.com/floyd). To simulate the winding resistance of the inductor, a 100 Ω resistor has been added in series in the computer simulations. The resistance of coils varies widely with the size of the coil and the wiring used to make the coil, so you may have found a much different resistance in the experiment. The frequency response has been plotted on the Bode plotter, a fictitious instrument but with characteristics similar to a spectrum analyzer.

Three of the four files contain a simulated "fault"; one has "no fault". The file with no fault is named EXP19-2-nf. You may want to open this file to compare your results with the computer simulation. Then open each of the files with faults. Use the simulated instruments to investigate the circuit and determine the problem. The following are the filenames for circuits with troubleshooting problems for this experiment.

EXP19-2-f1

    Fault: _____

EXP19-2-f2

    Fault: _____

EXP19-2-f3

    Fault: _____

Application Assignment 9

REFERENCE

Text, Chapter 9; Application Assignment: Putting Your Knowledge to Work

Step 1   Compare the PC board with the schematic. Do they agree?

Step 2   Test the input to amplifier board 1. If incorrect, specify the likely fault:

Step 3   Test the input to amplifier board 2. If incorrect, specify the likely fault:

Step 4   Test the input to amplifier board 3. If incorrect, specify the likely fault:

RELATED EXPERIMENT

MATERIALS NEEDED

Resistors:
    One 1.0 kΩ, two 10 kΩ
Capacitors:
    One 0.1 μF, 1.0 μF

DISCUSSION

The capacitor tests described in Experiment 18 can be conducted only on a capacitor that has been removed from the circuit under test. Usually there are other components that could account for a circuit failure; you need to have an idea of the reason for the failure before you randomly check parts. If a capacitor fails because it is open, it has no effect on the dc voltages but will not pass ac. If it fails because it is shorted, both dc and ac paths are affected. Other failures (such as the wrong size component) may produce a partial failure.

The circuit shown in Figure AA–9–1 is similar to the problem presented in the text. Capacitor C_1 represents a coupling capacitor and R_1 and R_2 set up the bias conditions needed for an amplifier. R_3

represents additional source resistance. Start by investigating the circuit when it is operating normally. Find the ac and dc voltage drops across each component. Then open C_1 and check circuit operation. Are there any changes to the dc voltages with the open capacitor? Then test the circuit with a short across C_1 (use a jumper). Finally, assume a capacitor that is too small was accidentally put in the circuit. Replace C_1 with a 0.1 μF capacitor and test the circuit. Table AA–9–1 is set up to record your data. Write a conclusion for your observations.

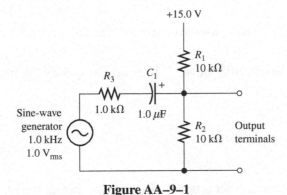

Figure AA–9–1

Table AA–9–1

Condition	Measured Voltages							
	V_{R1}		V_{R2}		V_{R3}		V_{C1}	
	dc	ac	dc	ac	dc	ac	dc	ac
Normal								
C_1 Open								
C_1 Shorted								
C_1 Wrong value								

EXPERIMENTAL RESULTS

164

Checkup 9

REFERENCE

Text, Chapter 9; Lab manual, Experiments 18 and 19

1. Assume two capacitors have the same voltage across them but capacitor A has twice the charge of capacitor B. From this we can conclude that:
   (a) A is larger.
   (b) They are equal.
   (c) B is larger.
   (d) No conclusion can be made.

2. Assume two capacitors have equal capacitances, but capacitor A has twice the voltage of capacitor B. From this we can conclude that:
   (a) A has larger plates.
   (b) A has smaller plates.
   (c) A has a greater charge.
   (d) A has less charge.

3. Assume capacitor A is larger than capacitor B. If they are connected in series, the total capacitance will be:
   (a) larger than A
   (b) smaller than B
   (c) larger than B
   (d) between A and B

4. Compared to any one capacitor, the total capacitance of three equal parallel capacitors is:
   (a) one-third   (b) the same   (c) double   (d) three times

5. Assume a 100 μF capacitor is charged to 10 V. The stored charge is:
   (a) 10 μC   (b) 100 μC   (c) 110 μC   (d) 1000 μC

6. In a series RC circuit, the time required for a capacitor to go from no charge to full charge (99%) is:
   (a) one time constant
   (b) three time constants
   (c) five time constants
   (d) 100 ms

7. A sinusoidal voltage waveform is applied to a capacitor. The amount of current is inversely proportional to the:
   (a) reactance   (b) capacitance   (c) frequency   (d) resistance

8. A sinusoidal voltage waveform is applied to a capacitor. If the frequency of the waveform is increased, the capacitance:
   (a) increases   (b) does not change   (c) decreases

9. The unit of measurement for capacitive reactance is the:
   (a) volt   (b) ohm   (c) farad   (d) coulomb

10. The power that is stored or returned to the circuit from a capacitor is called:
(a) stored power   (b) apparent power
(c) true power   (d) reactive power

11. The time constant of an RC circuit is measured with an oscilloscope and found to require 7.6 divisions to change from 0 to 63% of the final value. The SEC/DIV control is set to 20 μs/div.
(a) If the resistance is 4.7 kΩ, what is the measured value of the capacitance?

(b) How long after charging begins does it take the capacitor to reach full charge?

12. Assume you want to check a 100 μF capacitor to see if it is capable of storing a charge. What simple test would you perform?

13. Consider the circuit shown in Figure C–9–1, which is the same as Figure 19–2. You should have measured a larger voltage across C_1 than across R_1. What does this immediately tell you about the capacitive reactance at this frequency?

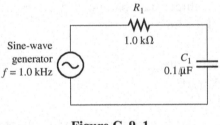

Figure C–9–1

14. Consider the circuit shown in Figure C–9–2. C_1 is known to be 0.047 μF, but the value of C_2 is unknown. Assume you measure 6.8 V$_{rms}$ across C_2. What is its capacitance?

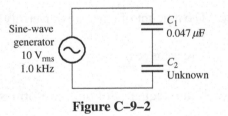

Figure C–9–2

15. A capacitor is marked 221. What is its value in pF and in μF?

20 Series *RC* Circuits

Name _____
Date _____
Class _____

READING
Text, Sections 10–1 through 10–3

OBJECTIVES
After performing this experiment, you will be able to:
1.   Compute the capacitive reactance of a capacitor from voltage measurements in a series *RC* circuit.
2.   Draw the impedance and voltage phasor diagrams for a series *RC* circuit.
3.   Explain how frequency affects the impedance and voltage phasors in a series *RC* circuit.

MATERIALS NEEDED
One 6.8 kΩ resistor
One 0.01 μF capacitor

SUMMARY OF THEORY
When a sine wave at some frequency drives a circuit that contains only linear elements (resistors, capacitors, and inductors), the waveforms throughout the circuit are also sine waves at that same frequency. To understand the relationship between the sinusoidal voltages and currents, we can represent ac waveforms as phasor quantities. A *phasor* is a complex number used to represent a sine wave's amplitude and phase. A graphical representation of the phasors in a circuit is a useful tool for visualizing the amplitude and phase relationship of the various waveforms. The algebra of complex numbers can then be used to perform arithmetic operations on sine waves.

Figure 20–1(a) shows an *RC* circuit with its impedance phasor diagram plotted in Figure 20–1(b). The total impedance is 5 kΩ, producing a current in this example of 1.0 mA. In any series circuit, the same current is throughout the circuit. By multiplying each of the phasors in the impedance diagram by the current in the circuit, we arrive at the voltage phasor diagram illustrated in Figure 20–1(c). It is convenient to use current as the reference for comparing voltage phasors because the current is the same throughout. Notice the direction of current. The voltage and the current are in the same direction across the resistor because they are in phase, but the voltage across the capacitor lags the current by 90°. The generator voltage is the phasor sum of the voltage across the resistor and the voltage across the capacitor.

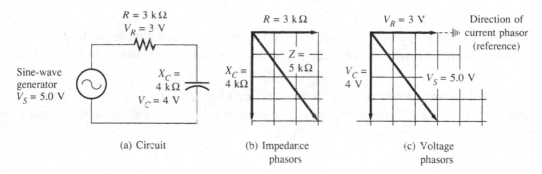

(a) Circuit   (b) Impedance phasors   (c) Voltage phasors

Figure 20–1

The phasor diagram illustrated by Figure 20–1 is correct only at one frequency. This is because the reactance of a capacitor is frequency dependent as given by the equation:

$$X_C = \frac{1}{2\pi f C}$$

As the frequency is raised, the reactance (X_C) of the capacitor decreases. This changes the phase angle and voltages across the components. These changes are investigated in this experiment.

PROCEDURE

1. Measure the actual capacitance of a 0.01 μF capacitor and a 6.8 kΩ resistor. Enter the measured values in Table 20–1. If you cannot measure the capacitor, use the listed value.

2. Connect the series *RC* circuit shown in Figure 20–2. Set the signal generator for a 500 Hz sine wave at 3.0 V_{pp}. The voltage should be measured with the circuit connected. Set the voltage with a voltmeter, and check both voltage and frequency with the oscilloscope. Record all voltages and currents throughout this experiment as peak-to-peak values.

Table 20–1

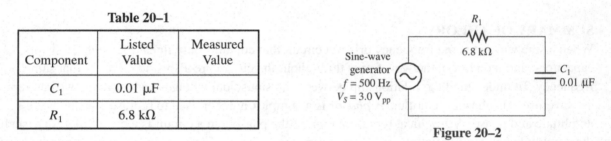

Component	Listed Value	Measured Value
C_1	0.01 μF	
R_1	6.8 kΩ	

Figure 20–2

3. Using the two-channel-difference technique described in Experiment 16, measure the peak-to-peak voltage across the resistor (V_R). Then measure the peak-to-peak voltage across the capacitor (V_C). Record the voltage readings on the first line of Table 20–2.

4. Compute the peak-to-peak current in the circuit by applying Ohm's law to the measured value of the resistor:

$$I = \frac{V_R}{R}$$

Since the current is the same throughout a series circuit, this is a simple method for finding the current in both the resistor and the capacitor. Enter this computed current in Table 20–2.

5. Compute the capacitive reactance, X_C, by applying Ohm's law to the capacitor. The reactance is found by dividing the voltage across the capacitor (step 3) by the current in the circuit (step 4). Enter the capacitive reactance in Table 20–2.

Table 20–2

Frequency	V_R	V_C	I	X_C	Z
500 Hz					
1000 Hz					
1500 Hz					
2000 Hz					
4000 Hz					
8000 Hz					

6. Compute the total impedance of the circuit by applying Ohm's law to the entire circuit. Use the generator voltage set in step 2 and the current determined in step 4. Enter the computed impedance in Table 20–2.

7. Change the frequency of the generator to 1000 Hz. Check the generator voltage and reset it to 3.0 V_{pp} if necessary. Repeat steps 3 through 6, entering the data in Table 20–2. Continue in this manner for each frequency listed in Table 20–2.

8. From the data in Table 20–2 and the measured value of R_1, draw the impedance phasors for the circuit at a frequency of 1000 Hz on Plot 20–1(a) and the voltage phasors on Plot 20–1(b).

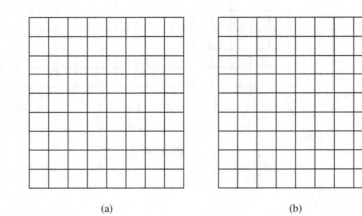

(a)   (b)

Plot 20–1

9. Repeat step 8 for a frequency of 4000 Hz. Draw the impedance phasors on Plot 20–2(a) and the voltage phasors on Plot 20–2(b).

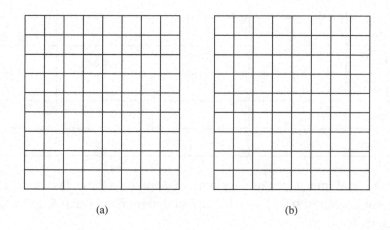

(a)   (b)

Plot 20–2

10. The phasor drawings reveal how the impedance and voltage phasors change with frequency. Investigate the frequency effect further by graphing both the voltage across the capacitor and the voltage across the resistor as a function of frequency. Label each curve. Use Plot 20–3.

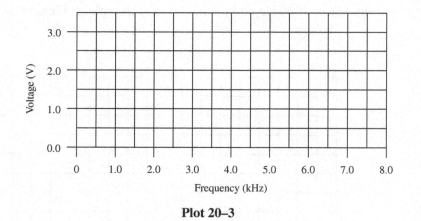

Plot 20–3

CONCLUSION

EVALUATION AND REVIEW QUESTIONS

1. The Pythagorean theorem can be applied to the phasors drawn in Plots 20–1 and 20–2. Show that the data in both plots satisfy the following equations

$$Z = \sqrt{R^2 + X_C^2}$$
$$V_S = \sqrt{V_R^2 + V_C^2}$$

2.   Assume you needed to pass high frequencies through an *RC* filter but block low frequencies. From the data in Plot 20–3, should you connect the output across the capacitor or across the resistor? Explain your answer.

3.   (a)   What happens to the total impedance of a series *RC* circuit as the frequency is increased?

        (b)   Explain why the phase angle between the generator voltage and the resistor voltage decreases as the frequency is increased.

4.   A student accidentally used a capacitor that was ten times larger than required in the experiment. Predict what happens to the frequency response shown in Plot 20–3 with the larger capacitor.

5.   Assume there was no current in the series *RC* circuit because of an open circuit. How could you quickly determine if the resistor or the capacitor were open?

FOR FURTHER INVESTIGATION

This experiment showed that the voltage phasor diagram can be obtained by multiplying each quantity on the impedance phasor diagram by the current in the circuit. In turn, if each of the voltage phasors is multiplied by the current, the resulting diagram is the power phasor diagram. Using the data from Table 20–2, convert the current and source voltage to an rms value. Then determine the true power, the reactive power, and the apparent power in the *RC* circuit at a frequency of 1000 Hz and a frequency of 4000 Hz. On Plot 20–4, draw the power phasor diagrams. (See Section 10–7 of the text for further discussion of the power phasors.)

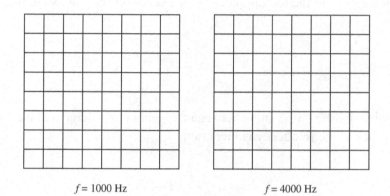

f = 1000 Hz $\qquad\qquad\qquad$ f = 4000 Hz

Plot 20–4

MULTISIM TROUBLESHOOTING

This experiment has four Multisim files on the website (www.prenhall.com/floyd). Three of the four files contain a simulated "fault"; one has "no fault". The file with no fault is named EXP20-2-nf. You may want to open this file to compare your results with the computer simulation. Then open each of the files with faults. Use the simulated instruments to investigate the circuit and determine the problem. The following are the filenames for circuits with troubleshooting problems for this experiment.

EXP20-2-f1

    Fault: _____

EXP20-2-f2

    Fault: _____

EXP20-2-f3

    Fault: _____

21 Parallel *RC* Circuits

READING
Text, Sections 10–4 through 10–9

OBJECTIVES
After performing this experiment, you will be able to:
1. Measure the current phasors for a parallel *RC* circuit.
2. Explain how the current phasors and phase angle are affected by a change in frequency for parallel *RC* circuits.

MATERIALS NEEDED
Resistors:
    One 100 kΩ, two 1.0 kΩ
Capacitors:
    One 1000 pF

SUMMARY OF THEORY
In a series circuit, the same *current* is in all components. For this reason, current is generally used as the reference. By contrast, in parallel, the same *voltage* is across all components. The voltage is therefore the reference. Current in each branch is compared to the circuit voltage. In parallel circuits, Kirchhoff's current law applies to any junction but care must be taken to add the currents as phasors. The current entering a junction is always equal to the current leaving the junction.

Figure 21–1 illustrates a parallel *RC* circuit. If the impedance of each branch is known, the current in that branch can be determined directly from Ohm's law. The current phasor diagram can then be constructed. The total current can be found as the phasor sum of the currents in each branch. The current in the capacitor is shown at +90° from the voltage reference because the current leads the voltage in a capacitor. The current in the resistor is along the *x*-axis because current and voltage are in phase in a resistor. The Pythagorean theorem can be applied to the current phasors, resulting in the equation

$$I_T = \sqrt{I_R^2 + I_C^2}$$

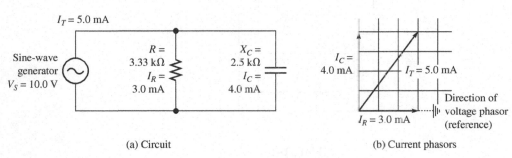

(a) Circuit   (b) Current phasors

Figure 21–1

In this experiment, two extra 1.0 kΩ resistors are added to "sense" current and provide a small voltage drop that can be measured. These resistors are much smaller than the parallel branch impedance, so their resistance can be ignored in the computation of circuit impedance.

PROCEDURE

1. Measure a resistor with a color-code value of 100 kΩ and each of two current-sense resistors (R_{S1} and R_{S2}) with color-code values of 1.0 kΩ. Measure the capacitance of a 1000 pF capacitor. Use the listed value if a measurement cannot be made. Record the measured values in Table 21–1.

Table 21–1   (f = 1.0 kHz)

	Listed Value	Measured Value	Voltage Drop	Computed Current
R_1	100 kΩ			
R_{S1}	1.0 kΩ			
R_{S2}	1.0 kΩ			
C_1	1000 pF			

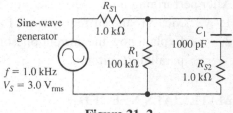

Figure 21–2

2. Construct the circuit shown in Figure 21–2. Set the generator to a voltage of 3.0 V_{rms} at 1.0 kHz. Check the voltage and frequency with your oscilloscope.

3. Using a voltmeter, measure the voltage drop across each resistor. The voltage drops are small, so measure as accurately as possible. You should keep three significant figures in your measurement. Record the voltage drops in Table 21–1.

4. Compute the current in each resistor using Ohm's law. Record the computed current in Table 21–1.

5. Draw the current phasors I_{R1}, I_{C1}, and the total current I_T on Plot 21–1. The total current is through sense resistor R_{S1}. The current I_{C1} is through sense resistor R_{S2}. Ignore the small effect of the sense resistors on the phasor diagram. Note carefully the direction of the phasors. Label each of the current phasors.

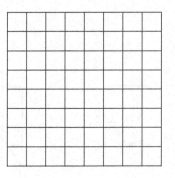

Plot 21–1

174

6. Compute X_{C1} for the 1.0 kHz frequency. Then, using this value and the measured resistance of R_1, find the total impedance, Z_T, of the circuit using the product-over-sum rule. The sense resistors can be ignored for this calculation.

$$X_{C1} = \underline{\hspace{2cm}} \qquad Z_T = \frac{R_1 X_C}{\sqrt{R_1^2 + X_C^2}} = \underline{\hspace{2.5cm}}$$

7. Using Z_T from step 6 and the applied voltage, V_S, compute the total current, I_T. The total current should basically agree with the value determined in step 4.

$$I_T = \underline{\hspace{2cm}}$$

8. Change the frequency of the generator to 2.0 kHz. Check that the generator voltage is still 3.0 V. Repeat steps 1–5 for the 2.0 kHz frequency. Enter the data in Table 21–2 and draw the current phasors on Plot 21–2.

Table 21–2 (f = 2.0 kHz)

	Listed Value	Measured Value	Voltage Drop	Computed Current
R_1	100 kΩ			
R_{S1}	1.0 kΩ			
R_{S2}	1.0 kΩ			
C_1	1000 pF			

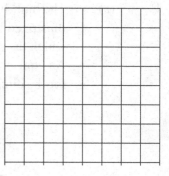

Plot 21–2

CONCLUSION

EVALUATION AND REVIEW QUESTIONS

1. Explain how increasing the frequency affects:
   (a) the total impedance of the circuit

   (b) the phase angle between the generator voltage and the generator current

2. Assume the frequency had been set to 5.0 kHz in this experiment. Compute:
    (a)   the current in the resistor

    (b)   the current in the capacitor

    (c)   the total current

3. If a smaller capacitor had been substituted in the experiment, what would happen to the current phasor diagrams?

4. (a) The high-frequency response of a transistor amplifier is limited by stray capacitance, as illustrated in Figure 21–3. The upper *cutoff* frequency is defined as the frequency at which the resistance R_{IN} is equal to the capacitive reactance X_C of the stray capacitance. An equivalent parallel *RC* circuit can simplify the problem. Compute the cutoff frequency for the circuit shown by setting $R_{IN} = X_C$ and solving for f_C.

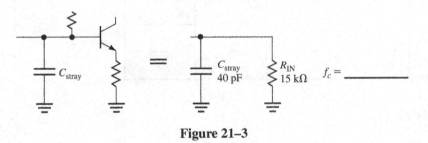

Figure 21–3

    (b)   How do the branch currents compare at the cutoff frequency?

    (c)   Explain what happens above this frequency to the current in the equivalent parallel *RC* circuit.

5. If the stray capacitance in Figure 21–3 is increased, what happens to the cutoff frequency?

FOR FURTHER INVESTIGATION

In a series *RC* circuit, the impedance phasor is the sum of the resistance and reactance phasors as shown in Experiment 20. In a parallel circuit, the admittance phasor is the sum of the conductance and the susceptance phasors. On Plot 21–3, draw the admittance, conductance, and susceptance phasors for the experiment at a frequency of 1.0 kHz. Hint: The admittance phasor diagram can be obtained directly from the current phasor diagram by dividing the current phasors by the applied voltage.

Plot 21–3

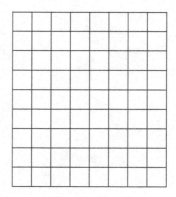

MULTISIM TROUBLESHOOTING

This experiment has four Multisim files on the website (www.prenhall.com/floyd). Three of the four files contain a simulated "fault"; one has "no fault". The file with no fault is named EXP21-2-nf. You may want to open this file to compare your results with the computer simulation. Then open each of the files with faults. Use the simulated instruments to investigate the circuit and determine the problem. The following are the filenames for circuits with troubleshooting problems for this experiment.

EXP21-2-f1

Fault: _____

EXP21-2-f2

Fault: _____

EXP21-2-f3

Fault: _____

Application Assignment 10

Name _____

Date _____

Class _____

REFERENCE

Text, Chapter 10; Application Assignment: Putting Your Knowledge to Work

Step 1   Evaluate the amplifier input circuit. Determine the equivalent resistance:
              Equivalent resistance = _____

Step 2   Measure the response at frequency f_1. The channel 1 response is shown. Sketch the
              waveform for channel 2 on Plot AA–10–1.

Step 3   Measure the response at frequency f_2. Sketch the waveform for channel 2 on
              Plot AA–10–2. Explain why the response is different than in step 2.

Step 4   Measure the response at frequency f_3. Sketch the waveform for channel 2 on
              Plot AA–10–3. Explain why the response is different from that in step 3.

Step 5   Plot a response curve for the amplifier input circuit on Plot AA–10–4.

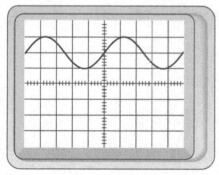

Plot AA–10–1   Ch1 and Ch2 = 0.5 V/div, 0.2 ms/div

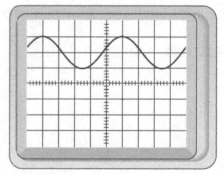

Plot AA–10–2   Ch1 and Ch2 = 0.5 V/div, 2.0 ms/div

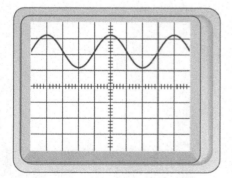

Plot AA–10–3   Ch1 and Ch2 = 0.5 V/div, 5.0 ms/div

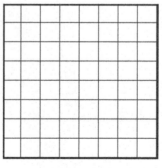

Plot AA–10–4

RELATED EXPERIMENT

MATERIALS NEEDED
One 100 kΩ resistor
One capacitor (value to be determined by student)

DISCUSSION
The application assignment requires you to consider the frequency response of a coupling capacitor. A circuit using a coupling capacitor was introduced in Application Assignment 9. The capacitor should look nearly like a short to the ac signal but appear open to the dc voltage. A simplified coupling circuit, with the dc portion removed, is illustrated in Figure AA–10–1. R_{input} represents the input resistance of an amplifier, and $C_{coupling}$ is the coupling capacitor.

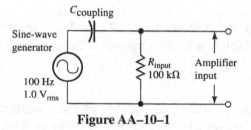

Figure AA–10–1

In this application, you need to find a capacitor that will allow a minimum of 90% of the generator signal to appear across R_{input} at a frequency of 100 Hz. Compute the value of a capacitor that will meet this requirement. Construct your circuit and test it by measuring the generator voltage, the voltage drop across the capacitor, and the voltage drop across the resistor using a 100 Hz signal from the generator. Summarize your calculations and measurements.

EXPERIMENTAL RESULTS

Checkup 10

REFERENCE

Text, Chapter 10; Lab manual, Experiments 20 and 21

1. If a sinusoidal voltage wave is applied to a capacitor, the current in the capacitor:
    (a) leads the voltage by 45°   (b) leads the voltage by 90°
    (c) lags the voltage by 45°   (d) lags the voltage by 90°

2. If a 0.1 µF capacitor is connected across a 50 V_{rms}, 1 kHz source, the current in the capacitor will be:
    (a) 2.00 nA   (b) 3.14 µA   (c) 0.795 mA   (d) 31.4 mA

3. In a series RC circuit in which $X_C = R$, the generator current:
    (a) leads the generator voltage by 45°   (b) leads the generator voltage by 90°
    (c) lags the generator voltage by 45°   (d) lags the generator voltage by 90°

4. In a parallel RC circuit in which $X_C = R$, the generator current:
    (a) leads the generator voltage by 45°   (b) leads the generator voltage by 90°
    (c) lags the generator voltage by 45°   (d) lags the generator voltage by 90°

5. If the frequency is raised in a series RC circuit and nothing else changes, the current in the circuit will:
    (a) increase   (b) stay the same   (c) decrease

6. If the frequency is raised in a parallel RC circuit and nothing else changes, the current in the circuit will:
    (a) increase   (b) stay the same   (c) decrease

7. The reciprocal of reactance is:
    (a) conductance   (b) susceptance   (c) admittance   (d) impedance

8. If the phase angle between the voltage and current in a series RC circuit is 30°, what is the power factor?
    (a) 0.5   (b) 0.707   (c) 0.866   (d) 1.0

9. In a purely capacitive circuit, the power factor is:
    (a) 0.0   (b) 0.5   (c) 0.707   (d) 1.0

10. To use a series RC circuit as a high-pass filter, the output should be taken from:
    (a) the resistor   (b) the capacitor   (c) the generator

11.   In a series circuit, the generator current is frequently used as the reference (see Experiment 20), but in a parallel circuit, the generator voltage is almost always used as the reference (see Experiment 21). Explain why.

12.   An *RC* circuit uses a 10 kΩ resistor in series with a 0.047 µF capacitor and is connected to a generator set to 10 V$_{rms}$.
   (a)   Determine the frequency at which $X_C = R$.

   (b)   At this frequency, what is the current in the circuit?

   (c)   At this frequency, what is the voltage across the resistor?

13.   In a certain parallel *RC* circuit, the generator current leads the generator voltage by 60°. Assume the generator is set to 10 V$_{rms}$ and the resistor has a value of 12.5 kΩ. Sketch the current phasor diagram on the Plot C–10–1. Label the values of currents on your diagram.

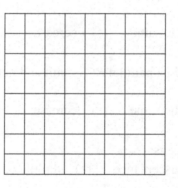

Plot C–10–1

14.   In Experiment 21, two 1.0 kΩ sense resistors were used to determine quickly the total current and the current in the capacitor. Would it be reasonable to use these same sense resistors for the problem circuit described in Question 13? Justify your answer.

READING
Text, Sections 11–1 through 11–4

OBJECTIVES
After performing this experiment, you will be able to:
1. Describe the effect of Lenz's law in a circuit.
2. Measure the time constant of an *LR* circuit and test the effects of series and parallel inductances on the time constant.

MATERIALS NEEDED
Two 7 H inductors (approximate value) (The secondary of a low-voltage transformer will work.
    The second inductor may be shared from another experiment.)
One neon bulb (NE-2 or equivalent)
One 33 kΩ resistor
For Further Investigation:
    One unknown inductor

SUMMARY OF THEORY
When there is a current through a coil of wire, a magnetic field is created around the wire. This electromagnetic field accompanies any moving electric charge and is proportional to the magnitude of the current. If the current changes, the electromagnetic field causes a voltage to be induced across the coil, which opposes the change. This property, which causes a voltage to oppose a change in current, is called *inductance*.

Inductance is the electrical equivalent of inertia in a mechanical system. It opposes a change in *current* in a manner similar to the way capacitance opposed a change in *voltage*. This property of inductance is described by Lenz's law. According to Lenz's law, an inductor develops a voltage across it that counters the effect of a *change* in current in the circuit. The induced voltage is equal to the inductance times the rate of change of current. Inductance is measured in *henries. One henry is defined as the quantity of inductance present when one volt is generated as a result of a current changing at the rate of one ampere per second.* Coils that are made to provide a specific amount of inductance are called *inductors*.

When inductors are connected in series, the total inductance is the sum of the individual inductors. This is similar to resistors connected in series. Likewise, the formula for parallel inductors is similar to the formula for parallel resistors. Unlike resistors, an additional effect can appear in inductive circuits. This effect is called *mutual inductance* and is caused by the interaction of the magnetic fields. The total inductance can be either increased or decreased due to mutual inductance.

Inductive circuits have a time constant associated with them, just as capacitive circuits do, except the rising exponential curve is a picture of the *current* in the circuit rather than the *voltage* as in the case of the capacitive circuit. Unlike the capacitive circuit, if the resistance is greater, the time constant is shorter. The time constant is found from the equation

$$\tau = \frac{L}{R}$$

where τ represents the time constant in seconds when *L* is in henries and *R* is in ohms.

PROCEDURE

1. In this step, you can observe the effect of Lenz's law. Connect the circuit shown in Figure 22–1 with a neon bulb in parallel with a large inductor. As noted in the materials list, the secondary of a low-voltage transformer can be used. Power should NOT be applied to the primary. Neon bulbs contain two insulated electrodes in a glass envelope containing neon gas. The gas will not conduct unless the voltage reaches approximately 70 V. When the gas conducts, the bulb will glow. When the switch is closed, dc current in the inductor is determined by the inductor's winding resistance. Close and open S_1 several times and observe the results.

   Observations:

2. Find out if the neon bulb will fire if the voltage is lowered. How low can you reduce the voltage source and still observe the bulb to glow? _____

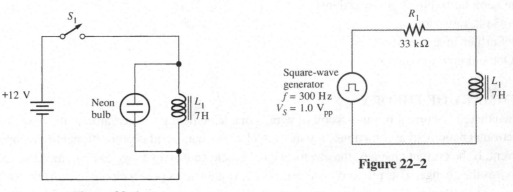

Figure 22–1   **Figure 22–2**

3. Connect the circuit shown in Figure 22–2. This circuit will be used to view the waveforms from a square wave generator. Set the generator, V_S, for a 1.0 V_{pp} square wave at a frequency of 300 Hz. This frequency is chosen to allow sufficient time to see the effects of the time constant. View the generator voltage on CH1 of a two-channel oscilloscope and the inductor waveform on CH2. If both channels are calibrated and have the VOLTS/DIV controls set to the same setting, you will be able to see the voltage across the resistor using the difference channel. Set the oscilloscope SEC/DIV control to 0.5 ms/div. Sketch the waveforms you see on Plot 22–1.

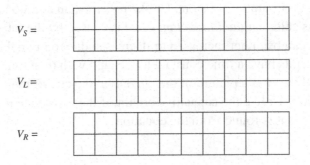

Plot 22–1

4.   Compute the time constant for the circuit. Enter the computed value in Table 22–1. Now measure the time constant by viewing the waveform across the resistor. The resistor voltage has the same shape as the current in the circuit, so you can measure the time constant by finding the time required for the resistor voltage to change from 0 to 63% of its final value. Stretch the waveform across the oscilloscope screen to make an accurate time measurement. Enter the measured time constant in Table 22–1.

Table 22–1

	Computed	Measured
Time constant, τ		

5.   When inductors are connected in series, the total inductance increases. When they are connected in parallel, the total inductance decreases. You can see the effect of decreasing the inductance by connecting a second 7 H inductor in parallel with the first. Note what happens to the voltage waveforms across the resistor and the inductor. Then connect the inductors in series and compare the effect on the waveforms. Describe your observations.

CONCLUSION

EVALUATION AND REVIEW QUESTIONS

1.   The ionizing voltage for a neon bulb is approximately 70 V. Explain how a 12 V source was able to cause the neon bulb to conduct.

2.   When a circuit containing an inductor is opened suddenly, an arc may occur across the switch. How does Lenz's law explain this?

3.   What is the total inductance when two 100 mH inductors are connected
         in series? _____   in parallel? _____

4.   What would happen to the time constant in Figure 22–2 if a 3.3 kΩ resistor were used instead of the 33 kΩ resistor?

5.   What effect does an increase in the frequency of the square wave generator have on the waveforms observed in Figure 22–2?

FOR FURTHER INVESTIGATION

Suggest a method in which you could use a square wave generator and a known resistor to determine the inductance of an unknown inductor. Then obtain an unknown inductor from your instructor and measure its inductance. Report on your method, results, and how your result compares to the accepted value for the inductor.

 **23 Inductive Reactance**

Name _____
Date _____
Class _____

READING
Text, Sections 11–5 and 11–6

OBJECTIVES
After performing this experiment, you will be able to:
1. Measure the inductive reactance of an inductor at a specified frequency.
2. Compare the reactance of inductors connected in series and parallel.

MATERIALS NEEDED
Two 100 mH inductors
One 1.0 kΩ resistor
For Further Investigation:
   One 12.6 V center-tapped transformer

SUMMARY OF THEORY
When a sine wave is applied to an inductor, a voltage is induced across the inductor as given by Lenz's law. When the *change* in current is a maximum, the largest induced voltage appears across the inductor. This is illustrated in Figure 23–1. Notice that when the current is not changing (at the peaks), the induced voltage is zero. For this reason, the voltage that appears across an inductor leads the current in the inductor by 1/4 cycle.

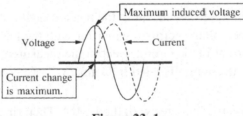

Figure 23–1

If we *raise* the frequency of the sine wave, the rate of change of current is increased and the value of the opposing voltage is increased. This results in a net *decrease* in the amount of current. Thus, the inductive reactance is increased by an increase in frequency. The inductive reactance is given by the equation

$$X_L = 2\pi f L$$

This equation reveals that a linear relationship exists between the inductance and the reactance at a constant frequency. Recall that in series, the total inductance is the sum of individual inductors (ignoring mutual inductance). The reactance of series inductors is, therefore, also the sum of the individual reactances. Likewise, in parallel, the reciprocal formula which applies to parallel resistors can be applied to both the inductance and the inductive reactance.

Ohm's law can be applied to inductive circuits. The reactance of an inductor can be found by dividing the voltage across the inductor by the current in it. That is,

$$X_L = \frac{V_L}{I_L}$$

PROCEDURE

1. Measure the inductance of each of two 100 mH inductors and record their measured values in Table 23–1. Measure and record the value of a 1.0 kΩ resistor. Use the listed values if you cannot measure the inductors.

2. Connect the circuit shown in Figure 23–2. Set the generator for a 1.0 kHz sine wave with a 1.0 V$_{rms}$. Measure the generator voltage with your DMM while it is connected to the circuit.[1] Check the frequency and voltage with the oscilloscope. Remember to convert the oscilloscope voltage reading to rms voltage to compare it to the DMM.

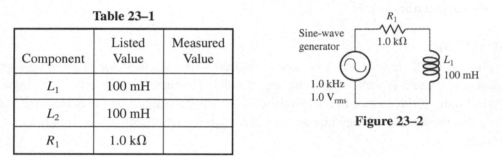

Table 23–1

Component	Listed Value	Measured Value
L_1	100 mH	
L_2	100 mH	
R_1	1.0 kΩ	

Figure 23–2

3. The circuit is a series circuit, so the current in the resistor is the identical current that is through the inductor. First, find the voltage across the resistor with the DMM. Then apply Ohm's law to the resistor to find the current in the circuit. Record the measured voltage and the computed current in Table 23–2 in the column labeled Inductor L_1.

4. Measure the voltage across the inductor with the DMM. Then find the inductive reactance by Ohm's law. Enter the values in Table 23–2.

5. Now compute the inductance based on the equation

$$L = \frac{X_L}{2\pi f}$$

Enter the computed inductance in Table 23–2.

---

[1]DMMs have a relatively low bandwidth, although most can measure 1.0 kHz. Verify that the DMM you are using has at least a 1.0 kHz bandwidth; if it does not, use the oscilloscope for all voltage measurements.

Table 23–2

	Inductor L_1	Inductor L_2
Voltage across R_1, V_R		
Total current, I		
Voltage across L, V_L		
Inductive reactance, X_L		
Computed inductance, L		

6. Replace L_1 with L_2 and repeat steps 3, 4, and 5. Enter the data in Table 23–2 in the column labeled Inductor L_2.

7. Place L_2 in series with L_1. Then find the inductive reactance for the series combination of the inductors as if they were one inductor. Enter the data in Table 23–3 in the column labeled Series Inductors. The following steps will guide you:
    (a) Check that the generator is set to 1.0 V rms. Find the current in the circuit by measuring the voltage across the resistor as before and dividing by the resistance.
    (b) Measure the voltage across *both* inductors.
    (c) Use Ohm's law to find the inductive reactance of both inductors. Use the voltage measured in step (b) and the current found in step (a).
    (d) Compute the total inductance by using the equation

$$L = \frac{X_L}{2\pi f}$$

8. Connect the inductors in parallel and repeat step 7. Assume the parallel inductors are one equivalent inductor for the measurements. Enter the data in Table 23–3 in the column labeled Parallel Inductors.

Table 23–3

Step		Series Inductors	Parallel Inductors
(a)	Voltage across R_1, V_R		
	Total current, I		
(b)	Voltage across inductors, V_L		
(c)	Inductive reactance, X_L		
(d)	Computed inductance, L		

CONCLUSION

EVALUATION AND REVIEW QUESTIONS
1.   (a)   Using the data in Table 23–2, compute the sum of the inductive reactances of the two inductors:

$$X_{L1} + X_{L2} =$$

    (b)   Using the data in Table 23–2, compute the product-over-sum of the inductive reactances of the two inductors:

$$\frac{(X_{L1})(X_{L2})}{X_{L1} + X_{L2}} =$$

    (c)   Compare the results from (a) and (b) with the reactances for the series and parallel connections listed in Table 23–3. What conclusion can you draw from these data?

2.   Repeat Question 1 using the data for the inductance, L. Compare the inductance of series and parallel inductors.

3.   What effect would an error in the frequency of the generator have on the data for this experiment?

4.   How could you apply the method used in this experiment to find the value of an unknown inductor?

5.   Compute the inductive reactance of a 50 μH inductor at a frequency of 50 MHz.

FOR FURTHER INVESTIGATION

A transformer consists of two or more coils wound on a common iron core. Frequently, one or more windings has a *center tap*, which splits a winding into two equal inductors. Because the windings are on the same core, mutual inductance exists between the windings. Obtain a small power transformer that has a low-voltage center-tapped secondary winding. Determine the inductance of each half of the winding using the method in this experiment. Then investigate what happens if the windings are connected in series. Keep the output of the signal generator constant for the measurements. Summarize your results.

MULTISIM TROUBLESHOOTING

This experiment has four Multisim files on the website (www.prenhall.com/floyd). Three of the four files contain a simulated "fault"; one has "no fault". The file with no fault is named EXP23-2-nf. You may want to open this file to compare your results with the computer simulation. Then open each of the files with faults. Use the simulated instruments to investigate the circuit and determine the problem. The following are the filenames for circuits with troubleshooting problems for this experiment.

EXP23-2-f1

   Fault: _____

EXP23-2-f2

   Fault: _____

EXP23-2-f3

   Fault: _____

Application
Assignment 11

REFERENCE

Text, Chapter 11; Application Assignment: Putting Your Knowledge to Work

Step 1   Measure the coil resistance and select a series resistor.

Step 2   Determine the time constant and the approximate inductance of coil 1.

Step 3   Determine the time constant and the approximate inductance of coil 2.

Step 4   Discuss how you could find the approximate inductance of the coils using a sinusoidal input instead of a square wave.

RELATED EXPERIMENT

MATERIALS NEEDED

Two decade resistance boxes
One 0.1 μF capacitor (for a standard)
One 1.0 kΩ resistor
One 100 mH indicator (or other value from about 1 mH to 100 mH)

DISCUSSION

A Maxwell bridge is commonly used to measure inductors that do not have a very high Q. It employs a fixed capacitor and two resistors as standards. The circuit for a Maxwell bridge is shown in Figure AA–11–1. Construct the bridge using two decade resistance boxes for R_1 and R_2 and a measured capacitor of 0.1 μF for C_1. A 100 mH inductor from Experiment 23 (or any unknown inductor from about 1 mH to 100 mH) can be used for the unknown. R_3 is a fixed 1.0 kΩ resistor. Measure the output voltage between terminals **A** and **B** with your DMM. Adjust both decade resistance boxes for the minimum voltage observed on the DMM.

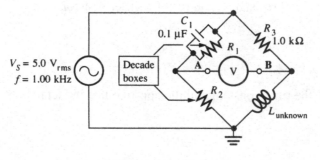

Figure AA–11–1

After you have adjusted the decade resistance boxes for minimum voltage across the **A** and **B** terminals, the circuit is a balanced ac bridge. The bridge is balanced when the product of the impedance of the diagonal elements is equal. The equations for the Maxwell bridge, given without proof, are

$$L_{unknown} = R_2 R_3 C_1$$

$$Q = 2\pi f C_1 R_1$$

Measure the unknown inductor with your Maxwell bridge. Compare your measurement with a laboratory bridge. What measurement errors account for the differences in the two measurements?

EXPERIMENTAL RESULTS

Checkup 11

Name _____
Date _____
Class _____

REFERENCE

Text, Chapter 11; Lab manual, Experiments 22 and 23

1. The voltage induced across an inductor by a changing magnetic field tends to:
    (a) oppose the current in the inductor   (b) oppose a change in the current
    (c) oppose the voltage across the inductor   (d) oppose a change in the voltage

2. Assume a neon bulb is in parallel with a large inductor. When current is interrupted by opening a switch, the neon bulb glows for a short time. This is due to:
    (a) the rapid change in resistance   (b) the time constant of the circuit
    (c) collapsing electric field   (d) induced voltage across the inductor

3. The total inductance of parallel inductors is always:
    (a) less than the smallest inductor   (b) less than the largest inductor
    (c) greater than the smallest inductor   (d) greater than the largest inductor

4. Assume two inductors have the same physical size and core material but inductor A has twice the number of windings of inductor B. From this we can conclude that the:
    (a) inductance of A is one-fourth that of B   (b) inductances are equal
    (c) inductance of A is twice that of B   (d) inductance of A is four times that of B

5. The time constant for a series RL circuit consisting of a 10 kΩ resistor and a 30 mH inductor is:
    (a) 3 μs   (b) 15 μs   (c) 30 μs   (d) 300 μs

6. The instant after the switch is closed in a series RL circuit, the voltage across the inductor is:
    (a) zero   (b) equal to the voltage across the resistor
    (c) 63% of the source voltage   (d) equal and opposite to the source voltage

7. The instant after the switch is closed in a series RL circuit, the voltage across the resistor is:
    (a) zero   (b) equal to the voltage across the inductor
    (c) 63% of the source voltage   (d) equal and opposite to the source voltage

8. One time constant after a switch is closed in a series RL circuit, the current will be:
    (a) 37% of its final value   (b) 50% of its final value
    (c) 63% of its final value   (d) 100% of its final value

9. The unit of inductive reactance is the:
    (a) farad   (b) henry   (c) ohm   (d) second

10. An inductor is connected across a sinusoidal generator. If the generator frequency is increased, the inductance:
(a) decreases   (b) stays the same   (c) increases

11. Name four factors that affect the inductance of a coil.

12. In Experiment 22, a large (7 H) inductor was used to "fire" the neon bulb. Why do you think a large inductor was specified? Would a smaller one work as well?

13. For a sinusoidal input, compare the phase difference between voltage and current for an *RC* circuit with that of an *RL* circuit.

14. The total inductance of two series inductors is 900 μH.
(a)   If one of the inductors is 350 μH, what is the inductance of the other?

(b)   Assume a 10 V sinusoidal waveform is applied to the two series inductors. What is the voltage across the 350 μH inductor?

15. In Experiment 22 (Figure 22–2), the frequency of the generator was set to 300 Hz. Could a higher frequency have been specified for this experiment? Why or why not?

16. For the circuit in Figure C–11–1, assume the voltage across the resistor is 2.1 V, and the voltage across the inductor is 4.0 V. If the source frequency is 100 kHz, determine the inductive reactance and the inductance of the unknown.

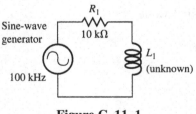

Figure C–11–1

24 Series *RL* Circuits

READING
Text, Sections 12–1 through 12–3

OBJECTIVES
After performing this experiment, you will be able to:
1.   Compute the inductive reactance of an inductor from voltage measurements in a series *RL* circuit.
2.   Draw the impedance and voltage phasor diagram for the series *RL* circuit.
3.   Measure the phase angle in a series circuit using either of two methods.

MATERIALS NEEDED
One 10 kΩ resistor
One 100 mH inductor

SUMMARY OF THEORY
When a sine wave drives a linear series circuit, the phase relationships between the current and the voltage are determined by the components in the circuit. The current and voltage are always in phase across resistors. With capacitors, the current is always leading the voltage by 90°, but for inductors, the voltage always leads the current by 90°. (A simple memory aid for this is *ELI the ICE man*, where *E* stands for voltage, *I* for current, and *L* and *C* for inductance and capacitance.)

Figure 24–1(a) illustrates a series *RL* circuit. The graphical representation of the phasors for this circuit is shown in Figure 24–1(b) and (c). As in the series *RC* circuit, the total impedance is obtained by adding the resistance and inductive reactance using the algebra for complex numbers. In this example, the current is 1.0 mA, and the total impedance is 5 kΩ. The current is the same in all components of a series circuit, so the current is drawn as a reference in the direction of the *x*-axis. If the current is multiplied by the impedance phasors, the voltage phasors are obtained as shown in Figure 24–1(c).

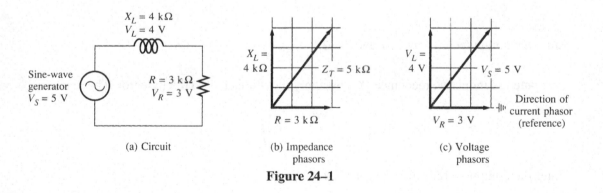

(a) Circuit   (b) Impedance phasors   (c) Voltage phasors

Figure 24–1

In this experiment, you learn how to make measurements of the phase angle. Actual inductors may have enough resistance to affect the phase angle in the circuit. You will use a series resistor that is large compared to the inductor's resistance to avoid this error.

PROCEDURE

1. Measure the actual resistance of a 10 kΩ resistor and the inductance of a 100 mH inductor. If the inductor cannot be measured, record the listed value. Record the measured values in Table 24–1.

2. Connect the circuit shown in Figure 24–2. Set the generator voltage with the circuit connected to 3.0 V_{pp} at a frequency of 25 kHz. The generator should have no dc offset. Measure the generator voltage and frequency with the oscilloscope as many meters cannot respond to the 25 kHz frequency. Use peak-to-peak readings for all voltage and current measurements in this experiment.

Table 24–1

Component	Listed Value	Measured Value
L_1	100 mH	
R_1	10 kΩ	

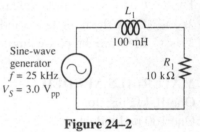

Sine-wave generator
f = 25 kHz
V_S = 3.0 V_{pp}

Figure 24–2

3. Using a two-channel oscilloscope, measure the peak-to-peak voltage across the resistor (V_R) and the peak-to-peak voltage across the inductor (V_L). (See Figure 24–3 for the setup.) Measure the voltage across the inductor using the difference technique described in Experiment 16. Record the voltage readings in Table 24–2.

Table 24–2 (f = 25 kHz)

V_R	V_L	I	X_L	Z_T

4. Compute the peak-to-peak current in the circuit by applying Ohm's law to the resistor. That is,

$$I = \frac{V_R}{R}$$

Enter the computed current in Table 24–2.

5. Compute the inductive reactance, X_L, by applying Ohm's law to the inductor. The reactance is

$$X_L = \frac{V_L}{I}$$

Enter the computed reactance in Table 24–2.

6. Calculate the total impedance (Z_T) by applying Ohm's law to the entire circuit. Use the generator voltage set in step 2 (V_S), and the current determined in step 4. Enter the computed impedance in Table 24–2.

7. Using the values listed in Tables 24–1 and 24–2, draw the impedance phasors on Plot 24–1(a) and the voltage phasors on Plot 24–1(b) for the circuit at a frequency of 25 kHz.

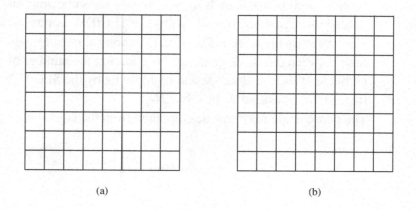

(a)   (b)

Plot 24–1

8. Compute the phase angle between V_R and V_S using the trigonometric relation

$$\theta = \tan^{-1}\left(\frac{V_L}{V_R}\right)$$

Enter the computed phase angle in Table 24–3.

9. Two methods for measuring phase angle will be explained. The first method can be used with any oscilloscope. The second can only be used with oscilloscopes that have a "fine" or variable SEC/DIV control. Measure the phase angle between V_R and V_S using one or both methods. The measured phase angle will be recorded in Table 24–3.

Phase Angle Measurement—Method 1
(a) Connect the oscilloscope so that channel 1 is across the generator and channel 2 is across the resistor. (See Figure 24–3.) Obtain a stable display showing between one and two cycles while viewing channel 1 (V_S). The scope should be triggered from channel 1.
(b) Measure the period, T, of the generator. Record it in Table 24–3. You will use this time in step (e).

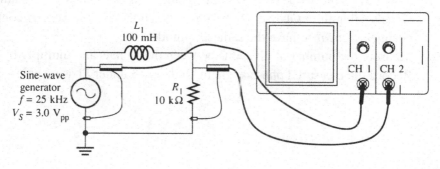

Figure 24–3

199

(c)   Set the oscilloscope to view both channels. (Do not have channel 2 inverted.) Adjust the amplitudes of the signals using the VOLTS/DIV, VERT POSITION, and the vernier controls until both channels *appear* to have the same amplitude as seen on the scope face.

(d)   Spread the signal horizontally using the SEC/DIV control until both signals are just visible across the screen. The SEC/DIV control must remain calibrated. Measure the time between the two signals, Δt, by counting the number of divisions along a horizontal graticule of the oscilloscope and multiplying by the SEC/DIV setting. (See Figure 24–4.) Record the measured Δt in Table 24–3.

(e)   The phase angle may now be computed from the equation

$$\theta = \left(\frac{\Delta t}{T}\right) \times 360°$$

Enter the measured phase angle in Table 24–3 under Phase Angle—Method 1.

Table 24–3

Computed Phase Angle θ	Measured Period T	Time Difference Δt	Phase Angle	
			Method 1 θ	Method 2 θ

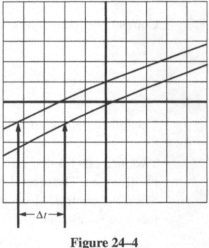

Figure 24–4

Phase Angle Measurement—Method 2

(a)   In this method the oscilloscope face will represent degrees, and the phase angle can be measured directly. The probes are connected as before. View channel 1 and obtain a stable display. Then adjust the SEC/DIV control and its vernier until you have exactly one cycle across the scope face. This is equivalent to 360° in 10 divisions, so each division is worth 36°.[1]

(b)   Now switch the scope to view both channels. As before, adjust the amplitudes of the signals using the VOLTS/DIV, VERT POSITION, and the vernier controls until both channels *appear* to have the same amplitude.

(c)   Measure the number of divisions between the signals and multiply by 36° per division. Record the measured phase angle in Table 24–3 under Phase Angle—Method 2.

---

[1]For even better resolution, you can set one-half cycle across the screen, making each division worth 18°. Care must be taken to center the waveform.

CONCLUSION

EVALUATION AND REVIEW QUESTIONS
1. (a) What will happen to the impedance in this experiment if the frequency increases?

    (b) What would happen to the impedance if the inductance were larger?

2. (a) What will happen to the phase angle in this experiment if the frequency increases?

    (b) What would happen to the phase angle if the inductance were larger?

3. Compute the percent difference between the computed phase angle and the method 1 phase angle measurement.

4. The critical frequency for an *RL* circuit occurs at the frequency at which the resistance is equal to the inductive reactance. That is, $R = X_L$. Since $X_L = 2\pi f L$ for an inductor, it can easily be shown that the circuit frequency for an *RL* circuit is

$$f_{crit} = \frac{R}{2\pi L}$$

Compute the critical frequency for this experiment. What is the phase angle between V_R and V_S at the critical frequency?

$f_{crit} =$ _____ $\theta =$ _____

5. A series *RL* circuit contains a 100 Ω resistor and a 1.0 H inductor and is operating at a frequency of 60 Hz. If 3.0 V are across the resistor, compute:

(a) the current in the inductor

(b) the inductive reactance, X_L

(c) the voltage across the inductor, V_L

(d) the source voltage, V_S

(e) the phase angle between V_R and V_S

FOR FURTHER INVESTIGATION

An older method for measuring phase angles involved interpreting Lissajous figures. A Lissajous figure is the pattern formed by the application of a sinusoidal waveform to both the *x*- and *y*-axes of an oscilloscope. Two signals of equal amplitude and exactly in phase will produce a 45° line on the scope face. If the signals are the same amplitude and exactly 90° apart, the waveform will appear as a circle. Other phase angles can be determined by applying the formula

$$\theta = \arcsin \frac{OA}{OB}$$

Figure 24–5 illustrates a Lissajous figure phase measurement. The measurements of *OA* and *OB* are along the *y*-axis. Try measuring the phase angle in this experiment using a Lissajous figure. You will have to have the signals the same amplitude and centered on the oscilloscope face. Then switch the time base of the oscilloscope to the XY mode.

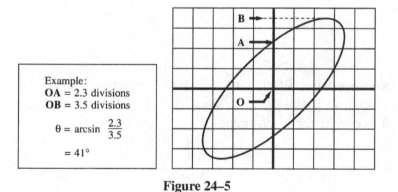

Example:
OA = 2.3 divisions
OB = 3.5 divisions

$\theta = \arcsin \dfrac{2.3}{3.5}$

= 41°

Figure 24–5

25 Parallel *RL* Circuits

Name _____
Date _____
Class _____

READING
Text, Sections 12–4 through 12–9

OBJECTIVES
After performing this experiment, you will be able to:
1.   Determine the current phasor diagram for a parallel *RL* circuit.
2.   Measure the phase angle between the current and voltage for a parallel *RL* circuit.
3.   Explain how an actual circuit differs from the ideal model of a circuit.

MATERIALS NEEDED
Resistors:
    One 3.3 kΩ, two 47 Ω
One 100 mH inductor

SUMMARY OF THEORY
The parallel *RC* circuit was investigated in Experiment 21. Recall that the circuit phasor diagram was drawn with current phasors and the voltage phasor was used as a reference, since voltage is the same across parallel components. In a parallel *RL* circuit, the current phasors will again be drawn with reference to the voltage phasor. The direction of the current phasor in a resistor is always in the direction of the voltage. Since current lags the voltage in an inductor, the current phasor is drawn at an angle of $-90°$ from the voltage reference. A parallel *RL* circuit and the associated phasors are shown in Figure 25–1.

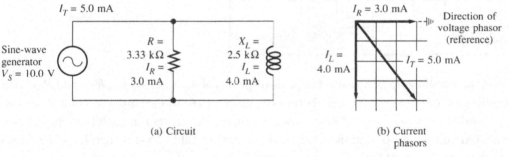

(a) Circuit   (b) Current phasors

Figure 25–1

    Practical inductors contain resistance that frequently is large enough to affect the purely reactive inductor phasor drawn in Figure 25–1. The resistance of an inductor can be thought of as a resistor in series with a pure inductor. The effect on the phasor diagram is to reduce an angle between I_L and I_R. In a practical circuit this angle will be slightly less than the $-90°$ shown in Figure 25–1. This experiment illustrates the difference between the approximations of circuit performance based on ideal components and the actual measured values.

    Recall that in Experiment 24, the phase angle between the source voltage, V_S, and the resistor voltage, V_R, in a series circuit were measured. The oscilloscope is a voltage-sensitive device, so comparing these voltages is straightforward. In parallel circuits, the phase angle of interest is usually between the total current, I_T, and one of the branch currents. To use the oscilloscope to measure the phase angle in a parallel circuit, we must convert the current to a voltage. This was done by inserting a small resistor in the branch where the current is to be measured. The resistor must be small enough not to have a major effect on the circuit.

PROCEDURE

1. Measure the actual resistance of a resistor with a color-code value of 3.3 kΩ and the resistance of two current-sensing resistors of 47 Ω each. Measure the inductance of a 100 mH inductor. Use the listed value if you cannot measure the inductor. Record the measured values in Table 25–1.

2. Measure the winding resistance of the inductor, R_W, with an ohmmeter. Record the resistance in Table 25–1.

3. Construct the circuit shown in Figure 25–2. Notice that the reference ground connection is at the low side of the generator. This connection will enable you to use a generator that does not have a "floating" common connection. Using your oscilloscope, set the generator to a voltage of 6.0 V_{pp} at 5.0 kHz. Check both the voltage and frequency with your oscilloscope. Record all voltages and currents in this experiment as peak-to-peak values.

Table 25–1

	Listed Value	Measured Value	Voltage Drop	Computed Current
R_1	3.3 kΩ			
R_{S1}	47 Ω			
R_{S2}	47 Ω			
L_1	100 mH			
R_W (L_1 resistance)				

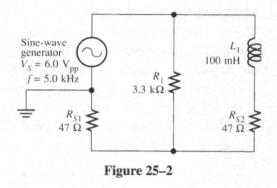

Figure 25–2

4. Using the oscilloscope, measure the peak-to-peak voltages across R_1, R_{S1}, and R_{S2}. Use the two-channel difference method (described in Experiment 16) to measure the voltage across the two ungrounded resistors. Apply Ohm's law to compute the current in each branch. Record the measured voltage drops and the computed currents in Table 25–1. Since L_1 is in series with R_{S2}, enter the same current for both.

5. The currents measured indirectly in step 4 are phasors because the current in the inductor is lagging the current in R_1 by 90°. The current in the inductor is the same as the current in R_{S2}, and the total current is through R_{S1}. Using the computed peak-to-peak currents from Table 25–1, draw the current phasors for the circuit on Plot 25–1. (Ignore the effects of the sense resistors.)

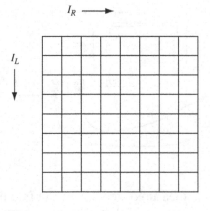

$I_R \longrightarrow$

I_L

Plot 25–1

Table 25–2

Phase Angle Between:	Computed	Measured
I_T and I_R		
I_R and I_L	90°	
I_T and I_L		

6. The phasor diagram illustrates the relationship between the total current and the current in each branch. Using the measured currents, compute the phase angle between the total current (I_T) and the current in R_1 (I_R). Then compute the phase angle between the total current (I_T), and the current in L_1 (I_L). Enter the computed phase angles in Table 25–2. (Note that the computed angles should add up to 90°, the angle between I_R and I_L.)

7. In this step, you will measure the phase angle between the generator voltage and current. This angle is approximately equal to the angle between I_T and I_R as shown in Figure 25–1. (Why?) Connect the oscilloscope probes as shown in Figure 25–3. Measure the phase angle using one of the methods in Experiment 24. The signal amplitudes in each channel are quite different, so the vertical sensitivity controls should be adjusted to make each signal appear to have the same amplitude on the scope. Record the measured angle between I_T and I_R in Table 25–2.

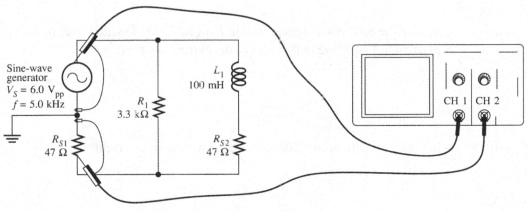

Figure 25–3

8. Replace R_{S1} with a jumper. This procedure enables you to reference the low side of R_1 and R_{S2}. Measure the angle between I_L and I_R by connecting the probes as shown in Figure 25–4. Ideally, this measurement should be 90°, but because of the coil resistance, you will likely find a smaller value. Adjust both channels for the same apparent amplitude on the scope face. Record your measured result in the second line in Table 25–2.

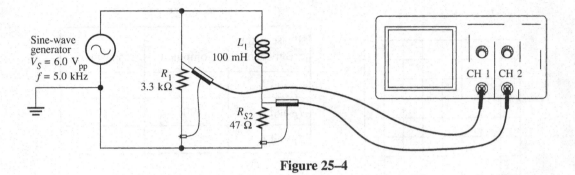

Figure 25–4

9. By subtracting the angle measured in step 7 from the angle measured step 8, you can find the phase angle between the I_T and I_L. Record this as the measured value on the third line of Table 25–2.

CONCLUSION

EVALUATION AND REVIEW QUESTIONS

1. If we assume that the currents determined in step 4 are 90° apart, the magnitude of the total current can be computed by applying the Pythagorean theorem to the current phasors. That is

$$I_T = \sqrt{I_{R1}^2 + I_{L1}^2}$$

(a) Compare the total current measured in R_{S1} (Table 25–1) with the current found by applying the Pythagorean theorem to the current phasors.

(b) What factors account for differences between the two currents?

2. How does the coil resistance measured in step 2 affect the angle between the current in the resistor and the current in the inductor?

3.   In Experiment 21, a 1.0 kΩ resistor was used as a current-sensing resistor. Why would this value be unsatisfactory in this experiment?

4.   If the inductor were open, what would happen to each?
       (a)   the total current in the circuit

       (b)   the phase angle between the generator voltage and current

       (c)   the generator voltage

5.   If the frequency were increased, what would happen to each?
       (a)   the total current in the circuit

       (b)   the phase angle between the generator voltage and current

       (c)   the generator voltage

FOR FURTHER INVESTIGATION

We could find the *magnitude* of the total current by observing the loading effect of the circuit on a signal generator. Consider the signal generator as a Thevenin circuit consisting of a zero impedance signal generator driving an internal series resistor consisting of the Thevenin source impedance. (See Figure 25–5.) When there is current to the external circuit, there is a voltage drop across the Thevenin resistance. The voltage drop across the Thevenin resistor, when divided by the Thevenin resistance, represents the total current in the circuit. To find the voltage drop across the Thevenin resistance, simply measure the difference in the generator voltage with the generator connected and disconnected from the circuit.

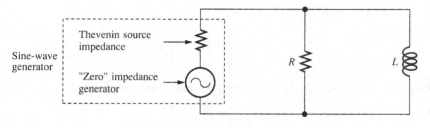

Figure 25–5

In addition to loading effects, the generator impedance also changes the phase angles in the circuit connected to it. If the impedance is smaller, the effect is greater. Investigate the loading effects for this experiment. Try finding the total current by the difference in loaded and unloaded voltage from the generator. What effect does the generator's impedance have on the phase angle?

Application
Assignment 12

Name _____
Date _____
Class _____

REFERENCE
Text, Chapter 12; Application Assignment: Putting Your Knowledge to Work

Step 1   From the resistance measurements of module 1, determine the arrangement of the two components and the values of the resistor and the winding resistance. Show the arrangement in the space provided.

Step 2   From the ac measurements of module 1, determine the value of the inductor indicated by the oscilloscope readings. Show your calculation.

Step 3   From the resistance measurements of module 2, determine the arrangement of the two components and the values of the resistor and the winding resistance. Show the arrangement in the space provided.

Step 4   From the ac measurements of module 2, determine the value of the inductor indicated by the oscilloscope readings. Show your calculation.

RELATED EXPERIMENT

MATERIALS NEEDED
One variable capacitor 12–100 pF (Mouser # ME242-3610-100 or equivalent)
One 100 mH inductor
One 27 kΩ resistor

DISCUSSION

A circuit will transfer maximum power to a load when the power factor is equal to 1. Maximum power factor is useful in certain impedance matching networks.

You can easily detect when the power factor is not maximum by observing a Lissajous figure on an oscilloscope. (See Experiment 24—For Further Investigation.) The phase angle is zero (power factor of 1) when the Lissajous figure shows a straight line on the oscilloscope display.

Construct the circuit shown in Figure AA–12–1. The oscilloscope is connected as shown. Adjust the VOLTS/DIV, VERT POSITION, and the vernier controls until both channels *appear* to have the same amplitude. Then, switch the oscilloscope to the XY mode. Measure and record the phase shift using the Lissajous figure as described in Experiment 24.

The results of the preceding measurement clearly show that the power factor is not 1. Add a variable capacitor of approximately 12–100 pF in series with the inductor, as shown in Figure AA–12–2. Observe the Lissajous figure and adjust the capacitor until the power factor is 1. Then, remove the capacitor and measure its value. Compare the reactance of the capacitor with the reactance of the inductor when the power factor is 1. What conclusion can you draw from this?

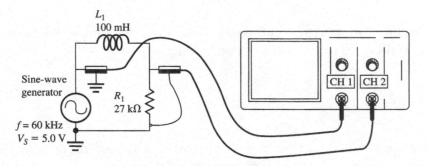

Figure AA–12–1

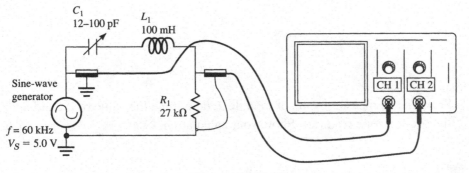

Figure AA–12–2

EXPERIMENTAL RESULTS

Checkup 12

REFERENCE

Text, Chapter 12; Lab manual, Experiments 24 and 25

1. If a sinusoidal voltage wave is applied to an inductor, the current in the inductor:
   (a) leads the voltage by 45°　　　　　(b) leads the voltage by 90°
   (c) lags the voltage by 45°　　　　　(d) lags the voltage by 90°

2. A 191 μH inductor is connected across a 20 V_{rms}, 50 MHz source. The current in the inductor will be approximately:
   (a) 333 μA　　(b) 3.3 mA　　(c) 33 mA　　(d) 333 mA

3. In a series RL circuit in which $X_L = R$, the generator current:
   (a) leads the generator voltage by 45°　　(b) leads the generator voltage by 90°
   (c) lags the generator voltage by 45°　　(d) lags the generator voltage by 90°

4. In a parallel RL circuit in which $X_L = R$, the generator current:
   (a) leads the generator voltage by 45°　　(b) leads the generator voltage by 90°
   (c) lags the generator voltage by 45°　　(d) lags the generator voltage by 90°

5. If the frequency is raised in a series RL circuit and nothing else changes, the current in the circuit:
   (a) increases　　(b) stays the same　　(c) decreases

6. If the frequency is raised in a parallel RL circuit and nothing else changes, the current in the circuit:
   (a) increases　　(b) stays the same　　(c) decreases

7. The admittance of a parallel RL circuit is 200 μS. If the total current is 400 μA, the applied voltage is:
   (a) 0.5　　(b) 2.0 V　　(c) 6.0 V　　(d) 8.0 V

8. A series RL circuit contains a 300 Ω resistor and an inductor with a reactance of 400 Ω. The total impedance of the circuit is:
   (a) 171 Ω　　(b) 350 Ω　　(c) 500 Ω　　(d) 700 Ω

9. A parallel RL circuit is connected to a 500 kHz voltage source of 16 V. The current in the inductor is 0.5 mA. The inductance is approximately:
   (a) 64 μH　　(b) 640 μH　　(c) 1.0 mH　　(d) 10 mH

10. In a certain series RL circuit, the phase angle is measured at 60° between the generator current and voltage. If the voltage across the inductor is 10 V, the voltage across the resistor is:
    (a) 5 V　　(b) 5.8 V　　(c) 8.66 V　　(d) 17.3 V

11. In Experiment 24 (step 9c), you were directed to make both signals appear to have the same amplitude on the scope face. Why is this best for minimizing measurement error?

12. In Experiment 25 (step 7), the statement is made that the phase angle between the generator voltage and current is approximately equal to the angle between I_T and I_R. Explain.

13. A parallel RL circuit is connected to a 10 V source. The current in the inductor has twice the magnitude of the current in the resistor. The total current is 7.0 mA.
    (a)   What is the phase angle between the generator current and voltage?

    (b)   Draw the current phasor diagram on Plot C–12–1.

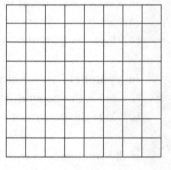

Plot C–12–1

14. For the circuit shown in Figure C–12–1, compute:
    (a)   the impedance seen by the generator _____
    (b)   total current from the generator _____
    (c)   voltage across L_1 _____
    (d)   phase angle between generator voltage and voltage across L_1 _____

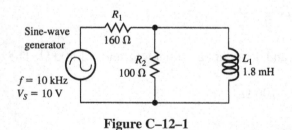

Figure C–12–1

26 Series Resonance

Name _____

Date _____

Class _____

READING
Text, Sections 13–1 through 13–4

OBJECTIVES
After performing this experiment, you will be able to:
1. Compute the resonant frequency, Q, and bandwidth of a series resonant circuit.
2. Measure the parameters listed in objective 1.
3. Explain the factors affecting the selectivity of a series resonant circuit.

MATERIALS NEEDED
Resistors:
> One 100 Ω, one 47 Ω

One 0.01 μF capacitor

One 100 mH inductor

SUMMARY OF THEORY
The reactance of inductors increases with frequency according to the equation

$$X_L = 2\pi f L$$

On the other hand, the reactance of capacitors decreases with frequency according to the equation

$$X_C = \frac{1}{2\pi f C}$$

Consider the series LC circuit shown in Figure 26–1(a). In any LC circuit, there is a frequency at which the inductive reactance is equal to the capacitive reactance. The point at which there is equal and opposite reactance is called *resonance*. By setting $X_L = X_C$, substituting the relations given above, and solving for f, it is easy to show that the resonant frequency of an LC circuit is

$$f_r = \frac{1}{2\pi\sqrt{LC}}$$

where f_r is the resonant frequency. Recall that reactance phasors for inductors and capacitors are drawn in opposite directions because of the opposite phase shift that occurs between inductors and capacitors. At series resonance these two phasors are added and cancel each other. This is illustrated in Figure 26–1(b). The current in the circuit is limited only by the total resistance of the circuit. The current in this example is 5.0 mA. If each of the impedance phasors is multiplied by this current, the result is the voltage phasor diagram as shown in Figure 26–1(c). Notice that the voltage across the inductor and the capacitor can be *greater* than the applied voltage!

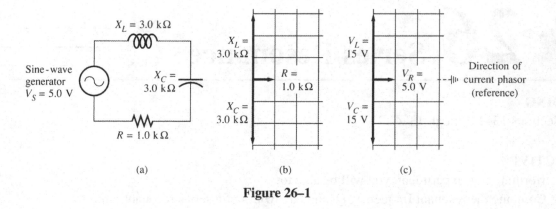

Figure 26–1

At the resonant frequency, the cancellation of the inductive and capacitive phasors leaves only the resistive phasor to limit the current in the circuit. Therefore, at resonance, the impedance of the circuit is a *minimum* and the current is a *maximum* and equal to V_S/R. The phase angle between the source voltage and current is zero. If the frequency is lowered, the inductive reactance will be smaller and the capacitive reactance will be larger. The circuit is said to be capacitive because the source current leads the source voltage. If the frequency is raised, the inductive reactance increases, and the capacitive reactance decreases. The circuit is said to be inductive.

The *selectivity* of a resonant circuit describes how the circuit responds to a group of frequencies. A highly selective circuit responds to a narrow group of frequencies and rejects other frequencies. The *bandwidth* of a resonant circuit is the frequency range at which the current is 70.7% of the maximum current. A highly selective circuit thus has a narrow bandwidth. The sharpness of the response to the frequencies is determined by the circuit Q. The Q for a series resonant circuit is the reactive power in either the coil or capacitor divided by the true power, which is dissipated in the total resistance of the circuit. The bandwidth and resonant frequency can be shown to be related to the circuit Q by the equation

$$Q = \frac{f_r}{BW}$$

Figure 26–2 illustrates how the bandwidth can change with Q. Responses 1 and 2 have the same resonant frequency but different bandwidths. The bandwidth for curve 1 is shown. Response curve 2 has a higher Q and a smaller BW. A useful equation that relates the circuit resistance, capacitance, and inductance to Q is

$$Q = \frac{1}{R}\sqrt{\frac{L}{C}}$$

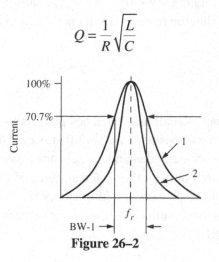

Figure 26–2

214

The value of R in this equation is the total equivalent series resistance in the circuit. Using this equation, the circuit response can be tailored to the application. For a highly selective circuit, the circuit resistance is held to a minimum and the L/C ratio is made high.

The Q of a resonant circuit can also be computed from the equation

$$Q = \frac{X_L}{R}$$

where X_L is the inductive reactance and R is again the total equivalent series resistance of the circuit. The result is the same if X_C is used in the equation, since the values are the same at resonance, but usually X_L is shown because the resistance of the inductor is frequently the dominant resistance of the circuit.

PROCEDURE

1. Measure the value of a 100 mH inductor, a 0.1 µF capacitor, a 100 Ω resistor, and a 47 Ω resistor. Enter the measured values in Table 26–1. If it is not possible to measure the inductor or capacitor, use the listed values.

2. Measure the winding resistance of the inductor, R_W. Enter the measured inductor resistance in Table 26–1.

Table 26–1

	Listed Value	Measured Value
L_1	100 mH	
C_1	0.01 µF	
R_1	100 Ω	
R_{S1}	47 Ω	
R_W (L_1 resistance)		

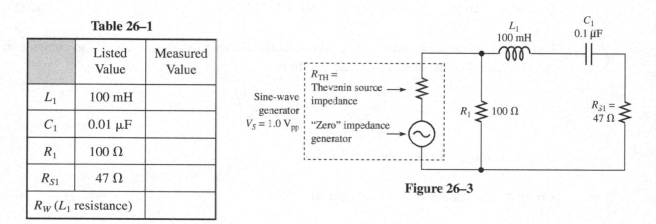

Figure 26–3

3. Construct the circuit shown in Figure 26–3. The purpose of the parallel 100 Ω resistor is to reduce the Thevenin driving impedance of the generator and, therefore, the total equivalent series resistance of the circuit.[1] Compute the total resistance of the equivalent series circuit. Note that looking back to the generator, R_{TH} is in parallel with R_1. In equation form, the equivalent series resistance, R_T, is

$$R_T = (R_{TH} \,|\, R_1) + R_W + R_{S1}$$

Enter the computed total resistance in Table 26–2.

---

[1]Some high-quality generators have a Thevenin resistance of 50 Ω. If you are using a 50 Ω generator, it is not necessary to include R_1.

Table 26–2

	Computed	Measured
R_T		
f_r		
Q		
V_{RS1}		
f_2		
f_1		
BW		

4. Using the measured values from Table 26–1, compute the resonant frequency of the circuit from the equation:

$$f_r = \frac{1}{2\pi\sqrt{LC}}$$

Record the computed resonant frequency in Table 26–2.

5. Use the total resistance computed in step 3 and the measured values of L and C to compute the approximate Q of the circuit from the equation:

$$Q = \frac{1}{R_T}\sqrt{\frac{L}{C}}$$

Enter the computed Q in Table 26–2.

6. Compute the bandwidth from the equation:

$$BW = \frac{f_r}{Q}$$

Enter this as the computed BW in Table 26–2.

7. Using your oscilloscope, tune for resonance by observing the voltage across the sense resistor, R_{S1}. As explained in the text, the current in the circuit rises to a maximum at resonance. The sense resistor will have the highest voltage across it at resonance. Measure the resonant frequency with the oscilloscope. Record the measured resonant frequency in Table 26–2.

8. Check that the voltage across R_1 is 1.0 V_{pp}. Measure the peak-to-peak voltage across the sense resistor at resonance. The voltage across R_{S1} is directly proportional to the current in the series LC branch, so it is not necessary to compute the current. Record in Table 26–2 the measured peak-to-peak voltage across R_{S1} (V_{RS1}).

9. Raise the frequency of the generator until the voltage across R_{S1} falls to 70.7% of the value read in step 7. Do not readjust the generator's amplitude in this step; this means that the Thevenin resistance of the generator is included in the measurement of the bandwidth. Measure and record this frequency as f_2 in Table 26–2.

10. Lower the frequency to below resonance until the voltage across R_{S1} falls to 70.7% of the value read in step 8. Again, do not adjust the generator amplitude. Measure and record this frequency as f_1 in Table 26–2.

11. Compute the bandwidth by subtracting f_1 from f_2. Enter this result in Table 26–2 as the measured bandwidth.

12. At resonance, the current in the circuit, the voltage across the capacitor, and the voltage across the inductor are all at a maximum value. Tune across resonance by observing the voltage across the capacitor, then try it on the inductor. Use the oscilloscope difference function technique described in Experiment 16. What is the maximum voltage observed on the capacitor? Is it the same or different than the maximum voltage across the inductor?

V_C (max) = _____ V_L (max) = _____

CONCLUSION

EVALUATION AND REVIEW QUESTIONS

1. (a) Compute the percent difference between the computed and measured bandwidth.

    (b) What factors account for the difference between the computed and measured values?

2. (a) What is the total impedance of the experimental circuit at resonance?_____

    (b) What is the phase shift between the total current and voltage?_____

3. (a) In step 12, you measured the maximum voltage across the capacitor and the inductor. The maximum voltage across either one should have been larger than the source voltage. How do you account for this?

    (b) Is this a valid technique for finding the resonant frequency? _____

4. (a) What happens to the resonant frequency, if the inductor is twice as large and the capacitor is half as large? _____

    (b) What happens to the bandwidth? _____

5. (a) Compute the resonant frequency for a circuit consisting of a 50 μH inductor in series with a 1000 pF capacitor. f_r = _____

    (b) If the total resistance of the above circuit is 10 Ω, what are Q and the bandwidth?
    Q = _____ BW = _____

FOR FURTHER INVESTIGATION

In this experiment, you measured three points on the response curve similar to Figure 26–1. Using the technique of measuring the voltage across R_{S1}, find several more points on the response curve. Graph your results on Plot 26–1.

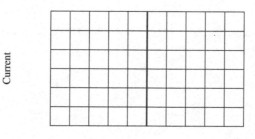

f_r
Frequency

Plot 26–1

MULTISIM TROUBLESHOOTING

This experiment has four Multisim files on the website (www.prenhall.com/floyd). Three of the four files contain a simulated "fault"; one has "no fault". The file with no fault is named EXP26-3-nf. You may want to open this file to compare your results with the computer simulation. Then open each of the files with faults. Use the simulated instruments to investigate the circuit and determine the problem. The following are the filenames for circuits with troubleshooting problems for this experiment.

EXP26-3-f1
    Fault: _____

EXP26-3-f2
    Fault: _____

EXP26-3-f3
    Fault: _____

READING
Text, Sections 13–5 through 13–8

OBJECTIVES
After performing this experiment, you will be able to:
1. Compute the resonant frequency, Q, and bandwidth of a parallel resonant circuit.
2. Measure the frequency response of a parallel resonant circuit.
3. Use the frequency response to determine the bandwidth of a parallel resonant circuit.

MATERIALS NEEDED
One 100 mH inductor
One 0.047 μF capacitor
One 1.0 kΩ resistor

SUMMARY OF THEORY
In an *RLC* parallel circuit, the current in each branch is determined by the applied voltage and the impedance of that branch. For an "ideal" inductor (no resistance), the branch impedance is X_L, and for a capacitor the branch impedance is X_C. Since X_L and X_C are functions of frequency, it is apparent that the currents in each branch are also dependent on the frequency. For any given L and C, there is a frequency at which the currents in each are equal and of opposite phase. This frequency is the resonant frequency and is found using the same equation as was used for series resonance:

$$fr = \frac{1}{2\pi\sqrt{LC}}$$

The circuit and phasor diagram for an ideal parallel *RLC* circuit at resonance is illustrated in Figure 27–1. Some interesting points to observe are: The total source current at resonance is equal to the current in the resistor. The total current is actually less than the current in either the inductor or the capacitor. This is because of the opposite phase shift which occurs between inductors and capacitors, causing the addition of the currents to cancel. Also, the impedance of the circuit is solely determined by

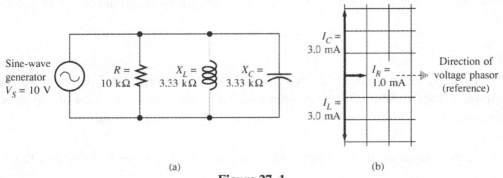

(a)

(b)

Figure 27–1

R, as the inductor and capacitor appear to be open. In a two-branch circuit consisting of only *L* and *C*, the source current would be zero, causing the impedance to be infinite! Of course, this does not happen with actual components that do have resistance and other effects.

In a practical two-branch parallel circuit consisting of an inductor and a capacitor, the only significant resistance is the winding resistance of the inductor. Figure 27–2(a) illustrates a practical parallel *LC* circuit containing winding resistance. By network theorems, the practical *LC* circuit can be converted to an equivalent parallel *RLC* circuit, as shown in Figure 27–2(b). The equivalent circuit is easier to analyze. The phasor diagram for the ideal parallel *RLC* circuit can then be applied to the equivalent circuit as was illustrated in Figure 27–1. The equations to convert the inductance and its winding resistance to an equivalent parallel circuit are

$$L_{eq} = L\left(\frac{Q^2 + 1}{Q^2}\right) \qquad R_{p(eq)} = R_W\,(Q^2 + 1)$$

where $R_{p(eq)}$ represents the parallel equivalent resistance, and R_W represents the winding resistance of the inductor. The *Q* used in the conversion equation is the *Q* for the inductor:

$$Q = \frac{X_L}{R_W}$$

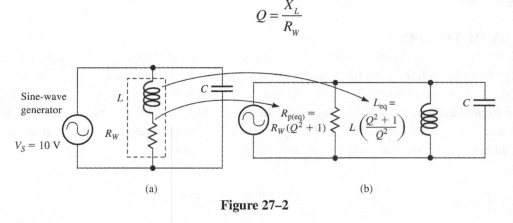

(a)   (b)

Figure 27–2

The *selectivity* of series circuits was discussed in Experiment 26. Parallel resonant circuits also respond to a group of frequencies. In parallel resonant circuits, the impedance as a function of frequency has the same shape as the current versus frequency curve for series resonant circuits. The *bandwidth* of a parallel resonant circuit is the frequency range at which the circuit impedance is 70.7% of the maximum impedance. The sharpness of the response to frequencies is again measured by the circuit *Q*. The circuit *Q* will be different from the *Q* of the inductor if there is additional resistance in the circuit. If there is no additional resistance in parallel with *L* and *C*, then the *Q* for a parallel resonant circuit is equal to the *Q* of the inductor.

PROCEDURE

1. Measure the value of a 100 mH inductor, a 0.047 μF capacitor, and a 1.0 kΩ resistor. Enter the measured values in Table 27–1. If it is not possible to measure the inductor or capacitor, use the listed values.

2. Measure the resistance of the inductor. Enter the measured inductor resistance in Table 27–1.

Table 27–1		
	Listed Value	Measured Value
L_1	100 mH	
C_1	0.047 μF	
R_{S1}	1.0 kΩ	
R_W (L_1 resistance)		

Table 27–2		
	Computed	Measured
f_r		
Q		
BW		

3.   Construct the circuit shown in Figure 27–3. The purpose of R_{S1} is to develop a voltage that can be used to sense the total current in the circuit. Compute the resonant frequency of the circuit using the equation

$$f_r = \frac{1}{2\pi\sqrt{LC}}$$

Enter the computed resonant frequency in Table 27–2. Set the generator to the f_r at 1.0 V_{pp} output, as measured with your oscilloscope. Use peak-to-peak values for all voltage measurements in this experiment.

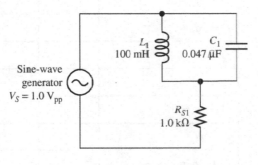

Figure 27–3

4.   The Q of a parallel LC circuit with no resistance other than the inductor winding resistance is equal to the Q of the inductor. Compute the approximate Q of the parallel LC circuit from

$$Q = \frac{X_L}{R_W}$$

Enter the computed Q in Table 27–2.

5.   Compute the bandwidth from the equation

$$BW = \frac{f_r}{Q}$$

Enter this as the computed BW in Table 27–2.

221

6. Connect your oscilloscope across R_{S1} and tune for resonance by observing the voltage across the sense resistor, R_{S1}. Resonance occurs when the voltage across R_{S1} is a minimum, since the impedance of the parallel LC circuit is highest. Measure the resonant frequency (f_r) and record the measured result in Table 27–2.

7. Compute a frequency increment (f_i) by dividing the computed bandwidth by 4. That is,

$$f_i = \frac{BW}{4}$$

Enter the computed f_i in Table 27–2.

8. Use the measured resonant frequency (f_r) and the frequency increment (f_i) from Table 27–2 to compute 11 frequencies according to the Computed Frequency column of Table 27–3. Enter the 11 frequencies in column 1 of Table 27–3.

Table 27–3

Computed Frequency	V_{RS1}	I	Z
$f_r - 5f_i =$			
$f_r - 4f_i =$			
$f_r - 3f_i =$			
$f_r - 2f_i =$			
$f_r - 1f_i =$			
$f_r =$			
$f_r + 1f_i =$			
$f_r + 2f_i =$			
$f_r + 3f_i =$			
$f_r + 4f_i =$			
$f_r + 5f_i =$			

9. Tune the generator to each of the computed frequencies listed in Table 27–3. At each frequency, check that the generator voltage is still at 1.0 V_{pp}; then measure the peak-to-peak voltage across R_{S1}. Record the voltage in column 2 of Table 27–3.

10. Compute the total peak-to-peak current, I, at each frequency by applying Ohm's law to the sense resistor R_{S1}. (That is, $I = V_{RS1}/R_{S1}$.) Record the current in column 3 of Table 27–3.

11. Use Ohm's law with the measured source voltage (1.0 V_{pp}) and source current at each frequency to compute the impedance at each frequency. Complete Table 27–3 by listing the computed impedances.

222

12. On Plot 27–1, draw the impedance versus frequency curve. From your curve determine the bandwidth. Complete Table 27–2 with the measured bandwidth.

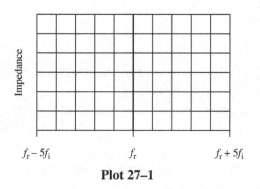

Plot 27–1

CONCLUSION

EVALUATION AND REVIEW QUESTIONS

1. (a) Compare the impedance as a function of frequency for series and parallel resonance.

   (b) Compare the current as a function of frequency for series and parallel resonance.

2. What was the phase shift between the total current and voltage at resonance?

3. At resonance the total current was a minimum, but the branch currents were not. How could you find the value of the current in each branch?

4.   What factors affect the Q of a parallel resonant circuit?

5.   In the circuit of Figure 27–2(a), assume the inductor is 100 mH with 120 Ω of winding resistance and the capacitor is 0.01 μF. Compute:

   (a)   the resonant frequency _____

   (b)   the reactance, X_L, of the inductor at resonance _____

   (c)   the Q of the circuit _____

   (d)   the bandwidth, BW _____

FOR FURTHER INVESTIGATION

The oscilloscope can be used to display the resonant dip in current by connecting a sweep generator to the circuit. This converts the time base on the oscilloscope to a frequency base. The sweep generator produces an FM (frequency modulated) signal, which is connected in place of the signal generator. In addition, the sweep generator has a synchronous sweep output that should be connected to the oscilloscope on the X channel input. The Y channel input is connected across the 1.0 kΩ sense resistor. The oscilloscope is placed in the XY mode. A diagram of the setup is shown in Figure 27–4. Build the circuit shown, determine a method to calibrate the frequency base, and summarize your procedure in a report.

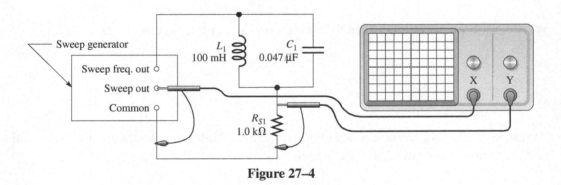

Figure 27–4

MULTISIM TROUBLESHOOTING

This experiment has four Multisim files on the website (www.prenhall.com/floyd). Three of the four files contain a simulated "fault"; one has "no fault". The file with no fault is named EXP27-3-nf. You may want to open this file to compare your results with the computer simulation. Then open each of the files with faults. Use the simulated instruments to investigate the circuit and determine the problem. The following are the filenames for circuits with troubleshooting problems for this experiment.

EXP27-3-f1

Fault: _____

EXP27-3-f2

Fault: _____

EXP27-3-f3

Fault: _____

 **Passive Filters**

READING
Text, Sections 13–4 through 13–7

OBJECTIVES
After performing this experiment, you will be able to:
1. Compare the characteristics and responses of low-pass, high-pass, bandpass, and notch filters.
2. Construct a T filter, a pi filter, and a resonant filter circuit and measure their frequency responses.

MATERIALS NEEDED
Resistors:
    One 680 Ω, one 1.6 kΩ
Capacitors:
    One 0.033 μF, two 0.1 μF
One 100 mH inductor

SUMMARY OF THEORY
In many circuits, different frequencies are present. If some frequencies are not desired, they can be rejected with special circuits called *filters*. Filters can be designed to pass either low or high frequencies. For example, in communication circuits, an audio frequency (AF) signal may be present with a radio frequency (RF) signal. The AF signal could be retained and the RF signal rejected with a *low-pass* filter. A *high-pass* filter will do the opposite: it will pass the RF signal and reject the AF signal. Sometimes the frequencies of interest are between other frequencies that are not desired. This is the case for a radio or television receiver, for example. The desired frequencies are present along with many other frequencies coming into the receiver. A resonant circuit is used to select the desired frequencies from the band of frequencies present. A circuit that passes only selected frequencies from a band is called a *bandpass* filter. The opposite of a bandpass filter is a *band reject* or *notch* filter. A typical application of a notch filter is to eliminate a specific interfering frequency from a band of desired frequencies. Figure 28–1 illustrates representative circuits and the frequency responses of various types of filters.

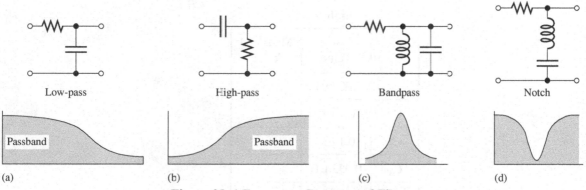

Low-pass   High-pass   Bandpass   Notch

Passband   Passband

(a)   (b)   (c)   (d)

Figure 28–1 Frequency Response of Filters

The simplest filters are *RC* and *RL* series circuits studied in Experiments 20 and 24. These circuits can be used as either high-pass or low-pass filters, depending on where the input and output voltages are applied and removed. A problem with simple *RC* and *RL* filters is that they change gradually from the passband to the stop band. You illustrated this characteristic on Plot 20–3 of Experiment 20.

Improved filter characteristics can be obtained by combining several filter sections. Unfortunately, you cannot simply stack identical sections together to improve the response as there are loading effects that must be taken into account. Two common improved filters are the *T* and the *pi* filters, so named because of the placement of the components in the circuit. Examples of T and pi filters are shown in Figure 28–2. Notice that the low-pass filters have inductors in series with the load and capacitors in parallel with the load. The high-pass filter is the opposite.

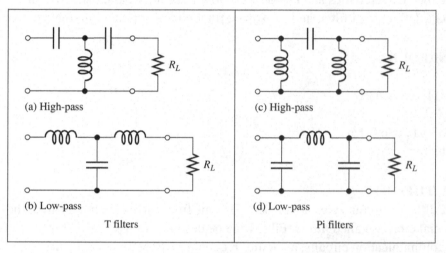

Figure 28–2

The choice of using a T or pi filter is determined by the load resistor and source impedance. If the load resistor is much larger than the source impedance, then the T-type filter is best. If the load resistor is much lower than the source impedance, then the pi filter is best.

PROCEDURE

1.   Obtain the components listed in Table 28–1. For this experiment, it is important to have values that are close to the listed ones. Measure all components and record the measured values in Table 28–1. Use listed values for those components that you cannot measure.

Table 28–1

	Listed Value	Measured Value
L_1	100 mH	
C_1	0.1 μF	
C_2	0.1 μF	
C_3	0.033 μF	
R_{L1}	680 Ω	
R_{L2}	1.6 kΩ	

2. Construct the pi filter circuit illustrated in Figure 28–3. Set the signal generator for a 500 Hz sine wave at 3.0 V$_{rms}$. The voltage should be measured at the generator with the circuit connected. Set the voltage with a voltmeter and check both voltage and frequency with the oscilloscope.

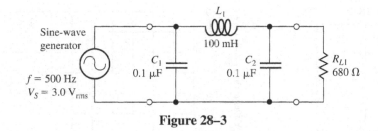

Figure 28–3

3. Measure and record the rms voltage across the load resistor (V_{RL1}) at 500 Hz. Record the measured voltage in Table 28–2.

4. Change the frequencies of the generator to 1000 Hz. Readjust the generator's amplitude to 3.0 V$_{rms}$. Measure V_{RL1}, and enter the data in Table 28–2. Continue in this manner for each frequency listed in Table 28–2. (Note: You may be unable to obtain 3.0 V from the generator at 8.0 kHz.)

5. Graph the voltage across the load resistor (V_{RL1}) as a function of frequency on Plot 28–1.

Table 28–2

Frequency	V_{RL1}
500 Hz	
1000 Hz	
1500 Hz	
2000 Hz	
3000 Hz	
4000 Hz	
8000 Hz	

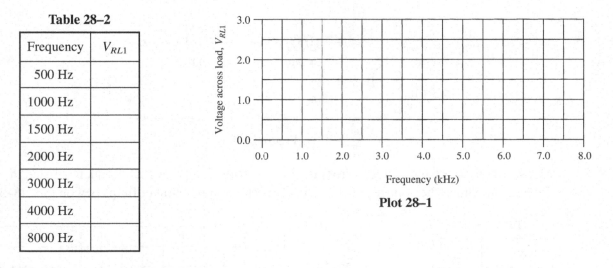

Plot 28–1

6. Construct the T filter circuit illustrated in Figure 28–4. Set the signal generator for a 500 Hz sine wave at 3.0 V$_{rms}$. The voltage should be measured with the circuit connected. Set the voltage with a voltmeter and check both voltage and frequency with the oscilloscope as before.

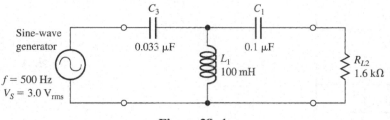

Figure 28–4

7. Measure and record the voltage across the load resistor (V_{RL2}) for each frequency listed in Table 28–3. Keep the generator voltage at 3.0 V_{rms}. Graph the voltage across the load resistor (V_{RL2}) as a function of frequency on Plot 28–2.

Table 28–3

Frequency	V_{RL2}
500 Hz	
1000 Hz	
1500 Hz	
2000 Hz	
3000 Hz	
4000 Hz	
8000 Hz	

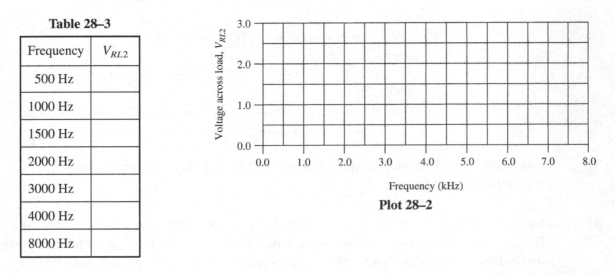

Plot 28–2

8. Construct the series resonant filter circuit illustrated in Figure 28–5. Set the generator for 3.0 V_{rms} at 500 Hz.

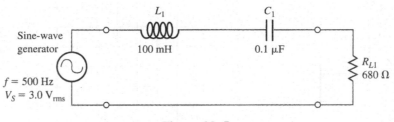

Figure 28–5

9. Measure and record the voltage across the load resistor (V_{RL1}) for each frequency listed in Table 28–4. Graph the voltage across the load resistor as a function of frequency on Plot 28–3.

Table 28–4

Frequency	V_{RL1}
500 Hz	
1000 Hz	
1500 Hz	
2000 Hz	
3000 Hz	
4000 Hz	
8000 Hz	

Plot 28–3

CONCLUSION

EVALUATION AND REVIEW QUESTIONS

1. The cutoff frequency for each filter in this experiment is that frequency at which the output is 70.7% of its maximum value. From the frequency response curves in Plots 28–1 and 28–2, estimate the cutoff frequency for the high- and low-pass filters.

   Pi filter cutoff frequency = _____

   T filter cutoff frequency = _____

2. Compare the response curve of the high and low filters in this experiment with the response curve from the simple *RC* filter in Experiment 20.

3. For each filter constructed in this experiment, identify it as a low-pass, high-pass, bandpass, or notch filter:

   (a)   Plot 28–1 (pi filter): _____

   (b)   Plot 28–2 (T filter): _____

   (c)   Plot 28–3 (resonant filter): _____

4. Explain what would happen to the response curve from the series resonant filter if the output were taken across the inductor and capacitor instead of the load resistor.

5. (a)   Sketch the circuit for a parallel resonant filter used as a bandpass filter.

   (b)   Sketch the circuit for a parallel resonant filter used as a notch filter.

FOR FURTHER INVESTIGATION

Using the components from this experiment, construct a parallel resonant notch filter. Measure the frequency response with a sufficient number of points to determine the bandwidth accurately. The bandwidth (*BW*) of a resonant filter is the difference in the two frequencies at which the response is 70.7% of the maximum output. From your data, determine the *BW* of the parallel notch resonant filter. Complete Table 28–5 and plot the data in Plot 28–4 for your notch filter. Write a conclusion.

Table 28–5

Frequency	V_{RL1}
500 Hz	
1000 Hz	
1500 Hz	
2000 Hz	
3000 Hz	
4000 Hz	
8000 Hz	

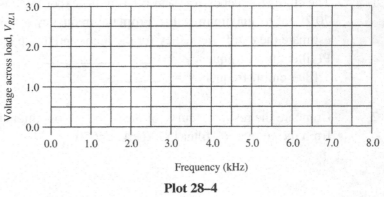

Plot 28–4

MULTISIM TROUBLESHOOTING

This experiment has four Multisim files on the website (www.prenhall.com/floyd). Three of the four files contain a simulated "fault"; one has "no fault". The file with no fault is named EXP28-3-nf. You may want to open this file to compare your results with the computer simulation. Then open each of the files with faults. Use the simulated instruments to investigate the circuit and determine the problem. The following are the filenames for circuits with troubleshooting problems for this experiment.

EXP28-3-f1

   Fault: _____

EXP28-3-f2

   Fault: _____

EXP28-3-f3

   Fault: _____

Application Assignment 13

Name _____

Date _____

Class _____

REFERENCE

Text, Chapter 13; Application Assignment: Putting Your Knowledge to Work

Step 1   From the oscilloscope displays shown in the text, plot the frequency response of the filter.

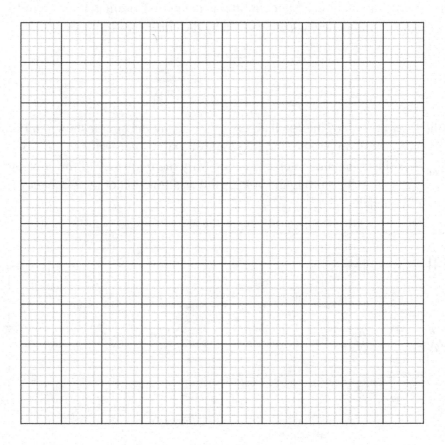

Plot AA–13–1

Step 2   Specify the type of filter and determine the resonant frequency and the bandwidth.

Type of filter is _____

Resonant frequency = _____   Bandwidth = _____

RELATED EXPERIMENT

MATERIALS NEEDED
Resistors:
 Two 10 kΩ, one 5.1 kΩ
Four 1000 pF capacitors

DISCUSSION
A bandstop, or notch, filter, as described in the text, is capable of removing certain undesired frequencies from a signal. Another application of a notch filter is the twin-T oscillator shown in Figure AA–13–1(a). It oscillates at the notch frequency, which is given by the equation

$$f_r = \frac{1}{2\pi RC}$$

 Test the filter portion of the oscillator by constructing the circuit shown in Figure AA–13–1(b). Use two 1000 pF capacitors in parallel for 2C. Connect a signal generator to the input and set it for a sine wave at 3.0 V_{pp} near the computed notch frequency. Vary the frequency of the generator and observe the output. Graph the response by plotting the voltage out as a function of the frequency for several points around the notch frequency.

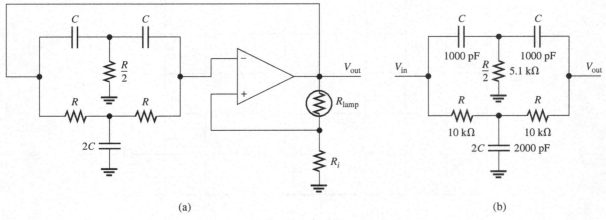

(a)   (b)

Figure AA–13–1

EXPERIMENTAL RESULTS

234

Checkup 13

REFERENCE

Text, Chapter 13; Lab manual, Experiments 26, 27, and 28

1. In a series resonant circuit at resonance:
    (a) current is a maximum
    (b) total impedance is zero
    (c) inductive reactance is larger than capacitive reactance
    (d) capacitive reactance is larger than inductive reactance

2. In a series *RLC* circuit, the phase angle between the capacitor voltage and inductor voltage is:
    (a) 0°   (b) 90°   (c) 180°   (d) dependent on the frequency

3. In a resonant circuit, if *L* is halved and *C* is doubled, the resonant frequency will:
    (a) remain the same   (b) double   (c) quadruple   (d) be halved

4. In a resonant circuit, if *L* is halved and *C* is doubled, *Q* will:
    (a) remain the same   (b) double   (c) be halved   (d) be one-fourth

5. In a parallel resonant circuit, if the frequency is higher than resonance, the circuit is said to be:
    (a) purely resistive   (b) inductive   (c) capacitive   (d) cutoff

6. At the cutoff frequency of a filter, the output voltage is approximately:
    (a) 10% of the input voltage   (b) 50% of the input voltage
    (c) 71% of the input voltage   (d) 100% of the input voltage

7. A bandstop filter can be made with a series resonant circuit and a resistor. The output is taken across:
    (a) the capacitor and the inductor   (b) the capacitor and resistor
    (c) the inductor and resistor   (d) the resistor

8. If the inductor in a resonant circuit is replaced with an identical inductor but the replacement inductor has higher coil resistance, the new bandwidth will be:
    (a) unchanged   (b) larger   (c) smaller

9. If a load is connected to a parallel resonant circuit, the selectivity will:
    (a) decrease   (b) remain the same   (c) increase

10. Assume a series *RLC* circuit is connected across a dc source. The dc voltage will be across:
    (a) the inductor   (b) the capacitor   (c) the resistor

11. A series *RLC* circuit has a 200 pF capacitor and a 100 μH inductor. If the inductor has a winding resistance of 20 Ω, calculate:
    (a) the resonant frequency

    (b) the impedance of the circuit at resonance

    (c) the *Q* of the coil

12. The series resonant circuit in Experiment 26 used a 100 Ω resistor in parallel with the source. Explain how this resistor affected the *Q* of the circuit.

13. A tank circuit is constructed using a 200 μH inductor and a variable capacitor. The circuit is required to tune the frequency range from 535 to 1605 kHz (AM radio band).
    (a) Compute the range of capacitance required to cause resonance over the range of frequencies.

    (b) Assuming the inductor has a resistance of 10 Ω, compute the *Q* of the circuit at each end of the tuning range.

    (c) Using the *Q* value found in (b), determine the bandwidth at each end of the tuning range.

14. Observe the parallel resonant circuit shown in Figure C–13–1. Assume the inductor has a winding resistance of 25 Ω. At resonance, calculate:
    (a) the inductive reactance

    (b) the *Q* of the coil

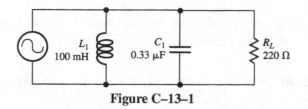

Figure C–13–1

236

29 Transformers

READING
Text, Sections 14–1 through 14–9

OBJECTIVES
After performing this experiment, you will be able to:
1. Determine the turns ratio for a transformer.
2. Show the phase relationships between the primary and secondary of a center-tapped transformer.
3. Compute the turns ratio required for matching a signal generator to a speaker.
4. Demonstrate how an impedance-matching transformer can increase the power transferred to a load.

MATERIALS NEEDED
One 12.6 V center-tapped transformer
One small impedance-matching transformer (approximately 600 Ω to 800 Ω)
One small speaker (4 or 8 Ω)
For Further Investigation
   One 100 Ω resistor

SUMMARY OF THEORY
A transformer consists of two (or more) closely coupled coils that share a common magnetic field. When an ac voltage is applied to the first coil, called the *primary,* a voltage is induced in the second coil, called the *secondary.* The voltage that appears across the secondary is proportional to the transformer turns ratio. The turns ratio is found by dividing the number of turns in the secondary winding by the number of turns in the primary winding. The turns ratio, *n,* is directly proportional to the primary and secondary voltages. That is,

$$n = \frac{N_S}{N_P} = \frac{V_S}{V_P}$$

For most work, we can assume that a transformer has no internal power dissipation and that all the magnetic flux lines in the primary also cut through the secondary—that is, we can assume the transformer is *ideal.* The ideal transformer delivers to the load 100% of the applied power. Actual transformers have losses due to magnetizing current, eddy currents, coil resistance, and so forth. In typical power applications, transformers are used to change the ac line voltage from one voltage to another or to isolate ac grounds. For the ideal transformer, the secondary voltage is found by multiplying the turns ratio by the applied primary voltage. That is,

$$V_S = nV_P$$

Since the ideal transformer has no internal losses, we can equate the power delivered to the primary to the power delivered by the secondary. Since $P = IV$, we can write:

$$\text{Power} = I_P V_P = I_S V_S$$

237

This equation shows that if the transformer causes the secondary voltage to be higher than the primary voltage, the secondary current must be less than the primary current. Also, if the transformer has no load, then there will be no primary or secondary current in the ideal transformer.

In addition to their ability to change voltages and isolate grounds, transformers are useful to change the resistance (or *impedance*) of a load as viewed from the primary side. (Impedance is a more generalized word meaning opposition to ac current.) The load resistance appears to increase by the turns ratio squared (n^2) when viewed from the primary side. Transformers used to change impedance are designed differently from power transformers. They need to transform voltages over a band of frequencies with low distortion. Special transformers called *audio,* or *wideband,* transformers are designed for this. To find the correct turns ratio needed to match a load impedance to a source impedance, use the following equation:

$$n = \sqrt{\frac{R_{load}}{R_{source}}}$$

In this experiment, you will examine both a power transformer and an impedance-matching transformer and calculate parameters for each.

PROCEDURE

1. Obtain a low-voltage power transformer with a center-tapped secondary (12.6 V secondary). Using an ohmmeter, measure the primary and secondary resistance. Record in Table 29–1.

2. Compute the turns ratio based on the normal line voltage (V_P) of 115 V and the specified secondary voltage of 12.6 V. Record this as the computed turns ratio, *n,* in Table 29–1.

3. For safety, we will use an audio generator in place of ac line voltages. Connect the circuit illustrated in Figure 29–1. Power transformers are designed to operate at a specific frequency (generally 60 Hz). Set the generator to a 60 Hz sine wave at 5.0 V$_{rms}$ on the primary. Measure the secondary voltage. From the measured voltages, compute the turns ratio for the transformer. Enter this value as the measured turns ratio in Table 29–1.

4. Compute the percent difference between the computed and measured turns ratio and enter the result in Table 29–1. The percent difference is found from the equation:

$$\%diff = \frac{n(meas.) - n(comp.)}{n(comp.)} \times 100\%$$

Table 29–1

Primary winding resistance, R_P	
Secondary winding resistance, R_S	
Turns ratio, n (computed)	
Turns ratio, n (measured)	
% difference	

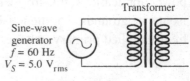

Sine-wave generator
$f = 60$ Hz
$V_S = 5.0$ V$_{rms}$

Transformer

Figure 29–1

5. Connect a two-channel oscilloscope to the secondary, as illustrated in Figure 29–2(a). Trigger the oscilloscope from channel 1. Compare the phase of the primary side viewed on channel 1 with the phase of the secondary side viewed on channel 2. Then reverse the leads on the secondary side. Describe your observations.

6. Connect the oscilloscope ground to the center tap of the transformer and view the signals on each side at the center tap at the same time as illustrated in Figure 29–2(b). Sketch the waveforms beside the figures showing measured voltages.

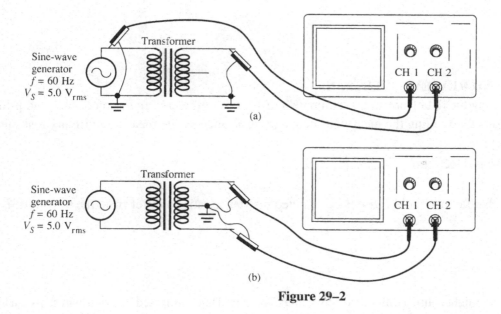

Figure 29–2

7. In this step, a transformer will be used to match a source impedance to a load impedance. A small speaker represents a low impedance (typically 4 or 8 Ω), whereas a signal generator is typically 600 Ω of Thevenin impedance. An impedance-matching transformer can make the load appear to have the same impedance as the source. This allows maximum power to be transferred to the load. Connect a small speaker directly to your signal generator and set the frequency to approximately 2 kHz. Note the volume of the sound from the speaker. Measure the voltage across the speaker.

$V_{SPKR} = $ _____

8. Using the specified Thevenin impedance of the generator and the specified speaker impedance, compute the turns ratio required to match the speaker with your signal generator.

$n = $ _____

9. Connect a small impedance-matching transformer into the circuit. It is not necessary to obtain the precise turns ratio required to note the improvement in the power delivered to the speaker. You can find the correct leads to the primary and secondary of the impedance-matching transformer using an ohmmeter. Since the required transformer is a step down type, the primary resistance will be higher than the secondary resistance. Often, the primary winding will have a center tap for push-pull amplifiers. Again measure the voltage across the speaker.

$V_{\text{SPKR}} =$ _____

CONCLUSION

EVALUATION AND REVIEW QUESTIONS

1. (a) Using the data from step 1, compute a resistance ratio between the secondary and primary coils by dividing the measured secondary resistance by the measured primary resistance.

Resistance ratio = _____

(b) What factors could cause the computed resistance ratio to differ from the turns ratio?

2. What factors might cause a difference between the measured and computed turns ratio in steps 2 and 3?

3. Compare the voltage across the speaker as measured in step 7 and in step 9. Explain why there is a difference.

4. The power supplied to an ideal transformer should be zero if there is no load. Why?

5.　(a)　If an ideal transformer has 115 V across the primary and draws 200 mA of current, what power is dissipated in the load?

　(b)　If the secondary voltage in the transformer of part (a) is 24 V, what is the secondary current?

　(c)　What is the turns ratio?

FOR FURTHER INVESTIGATION

The ideal transformer model neglects a small current that is in the primary independent of secondary load current. This current, called the *magnetizing* current, is required to produce the magnetic flux and is added to the current that is present due to the load. The magnetizing current appears to be through an equivalent inductor parallel to the ideal transformer. Investigate this current by connecting the circuit in Figure 29–3 using the impedance-matching transformer. Calculate the magnetizing current, I_M, in the primary by measuring the voltage across a series resistor with no load and applying Ohm's law:

$$I_M = \frac{V_R}{R}$$

Find out if the magnetizing current changes as frequency is changed. Be sure to keep the generator at a constant 5.0 V$_{rms}$.

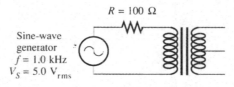

Figure 29–3

Application Assignment 14

Name _____
Date _____
Class _____

REFERENCE

Text, Chapter 14; Application Assignment: Putting Your Knowledge to Work

Step 1   Familiarization with the power supply.

Step 2   Measure voltages on power supply board 1. Determine from the readings whether or not the board is working properly. If not, isolate the problem to one of the items in Table AA–14–1.

Table AA–14–1

Board 1 is:
A. working properly
B. has the following problem:
(a)   rectifier, filter, or regulator
(b)   transformer
(c)   fuse
(d)   power source

Step 3   Measure voltages on power supply boards 2, 3, 4. Determine from the readings whether or not the board is working properly. If not, isolate the problem to one of the items in Table AA–14–2.

Table AA–14–2

Board 2 is:	Board 3 is:	Board 4 is:
A. working properly	A. working properly	A. working properly
B. has the following problem:	B. has the following problem:	B. has the following problem:
(a) rectifier, filter, or regulator	(a) rectifier, filter, or regulator	(a) rectifier, filter, or regulator
(b) transformer	(b) transformer	(b) transformer
(c) fuse	(c) fuse	(c) fuse
(d) power source	(d) power source	(d) power source

RELATED EXPERIMENT

MATERIALS NEEDED

One small speaker (4 or 8 Ω)

One 20 Ω variable resistor (approximate value)

DISCUSSION

To match the impedance of a speaker to an amplifier, you need to know the impedance of the speaker. You can measure the impedance of a speaker by the circuit shown in Figure AA–14–1. Set the scope controls as follows: CH 1, 0.1 V/div; CH 2, 50 mV/div (this is a 2:1 ratio); and SEC/DIV, 0.5 ms/div. The variable controls should be in the calibrated position. Adjust the peak-to-peak amplitude of the function generator to about 300 mV$_{pp}$ (actual value is not critical) and center both traces. Then adjust the potentiometer until the two signals appear as one—they should appear superimposed on each other. At this point, the impedance of the speaker is the same as the impedance of the potentiometer. (Why?) Remove the potentiometer and measure its resistance.

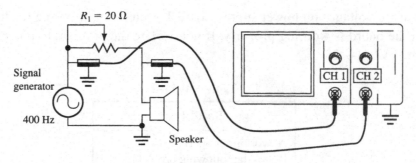

Figure AA–14–1

A variation of the method is to observe the waveform using the XY mode. Switch to XY mode but do not change the vertical sensitivity of the channels. At 400 Hz, the speaker impedance is primarily resistive, and a straight line at a 45° slope should be observed. Note how changing the potentiometer affects the line. Now try raising the frequency. Does the impedance of the speaker change? Try 20 kHz and see what happens. Summarize your findings.

EXPERIMENTAL RESULTS

Checkup 14

REFERENCE

Text, Chapter 14; Lab manual, Experiment 29

1. Transformers work by the principle of:
   - (a) self-inductance
   - (b) mutual inductance
   - (c) hysteresis
   - (d) coupled electric fields

2. The efficiency of an ideal transformer is:
   - (a) 90%
   - (b) 95%
   - (c) 100%
   - (d) dependent on the transformer

3. Air core transformers are primarily used for:
   - (a) impedance matching
   - (b) isolation
   - (c) power
   - (d) radio frequencies

4. The impedance seen on the primary side of an impedance-matching transformer is called:
   - (a) load resistance
   - (b) primary resistance
   - (c) reflected resistance
   - (d) winding resistance

5. A transformer with a single winding that can be adjusted with a sliding mechanism is known as a(n):
   - (a) variac
   - (b) rheostat
   - (c) isolation transformer
   - (d) tapped transformer

6. If a transformer has a much lower secondary voltage than primary voltage, which of the following is true?
   - (a) $P_S > P_P$
   - (b) $N_S > N_P$
   - (c) $I_S > I_P$
   - (d) efficiency is very poor

7. A transformer with a turns ratio of 2 has a primary voltage of 120 V. The secondary voltage is:
   - (a) 60 V
   - (b) 120 V
   - (c) 240 V
   - (d) 480 V

8. An ideal transformer with a turns ratio of 5 has a primary voltage of 120 V and a secondary load consisting of a 100 Ω resistor. The primary current is:
   - (a) 0.17 A
   - (b) 1.2 A
   - (c) 6.0 A
   - (d) 27.5 A

9. An impedance-matching transformer is needed to match an 8 Ω load to a 600 Ω source. The ideal reflected resistance is:
   - (a) 8 Ω
   - (b) 16 Ω
   - (c) 600 Ω
   - (d) 1200 Ω

10. The turns ratio for the transformer in Question 9 is:
    - (a) 0.0133
    - (b) 0.115
    - (c) 8.66
    - (d) 75

11.   In Experiment 29, you tested an impedance-matching transformer and a power transformer. Which transformer do you think is closest to the ideal transformer? Why?

12.   A power transformer with a primary voltage of 120 V has a secondary voltage of 28 V connected to a 47 Ω load resistor. If the primary power is 18 W, calculate:
       (a)   the efficiency

       (b)   the primary current

       (c)   the turns ratio

13.   Explain the difference between a tapped transformer and a multiple-winding transformer.

14.   What is the purpose of an isolation transformer?

15.   Assume you need to replace a missing fuse that is in series with the primary winding of a power transformer. The primary is designed for a 120 V, and the secondary is rated for 12.6 V at 1.0 A. What size fuse should you use? Justify your answer.

16.   An amplifier with a 50 Ω Thevenin impedance is used to drive a speaker with an 8 Ω impedance. Assume the amplifier has an unloaded output voltage of 10 V_{rms}.
       (a)   Compute the power delivered to the speaker with no impedance-matching transformer.

       (b)   Calculate the turns ratio of the impedance-matching transformer needed to maximize the power transfer.

READING
Text, Sections 15–1 through 15–9

OBJECTIVES
After performing this experiment, you will be able to:
1. Explain how an *RC* or *RL* series circuit can integrate or differentiate a signal.
2. Compare the waveforms for *RC* and *RL* circuits driven by a square wave generator.
3. Determine the effect of a frequency change for pulsed *RC* and *RL* circuits.

MATERIALS NEEDED
One 10 kΩ resistor
Capacitors:
    One 0.01 μF, one 1000 pF
One 100 mH inductor

SUMMARY OF THEORY
In mathematics, the word *integrate* means to sum. If we kept a running sum of the area under a horizontal straight line, the area would increase linearly. An example is the speed of a car. Let's say the car is traveling a constant 40 miles per hour. In 1/2 hour the car has traveled 20 miles. In 1 hour the car has traveled 40 miles, and so forth. The car's rate is illustrated in Figure 30–1(a). Each of the three areas shown under the rate curve represents 20 miles. The area increases linearly with time and is shown in Figure 30–1(b). This graph represents the integral of the rate curve.

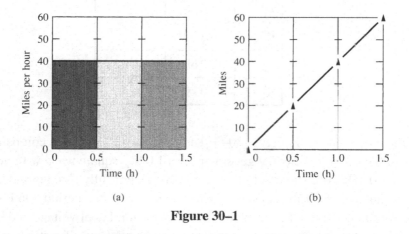

(a)   (b)

Figure 30–1

A similar situation exists when a capacitor starts to charge. If the applied voltage is a constant, the voltage on the capacitor rises exponentially. However, if we examine the beginning of this exponential rise, it appears to rise in a linear fashion. As long as the voltage change across the capacitor is small compared to the final voltage, the output will represent integration. An *integrator* is any circuit in which the output is proportional to the integral of the input signal. *If the RC time constant of the circuit is long compared to the period of the input waveform, then the waveform across the capacitor is integrated.*

247

The opposite of integration is *differentiation.* Differentiation means rate of change. *If the RC time constant of the circuit is short compared to the period of the input waveform, then the waveform across the resistor is differentiated.* A pulse waveform that is differentiated produces spikes at the leading and trailing edge as shown in Figure 30–2. Differentiator circuits can be used to detect the leading or trailing edge of a pulse. Diodes can be used to remove either the positive or negative spike.

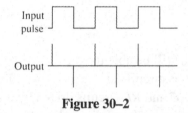

Figure 30–2

An *RL* circuit can also be used as an integrator or differentiator. As in the *RC* circuit, the time constant for the *RL* integrating circuit must be long compared to the period of the input waveform, and the time constant for the differentiator circuit must be short compared to the input waveform. The *RL* circuit will have similar waveforms to the *RC* circuit except that the output signal is taken across the inductor for the differentiating circuit and across the resistor for the integrating circuit.

PROCEDURE

1. Measure the value of a 100 mH inductor, a 0.01 μF and a 1000 pF capacitor, and a 10 kΩ resistor. Record their values in Table 30–1. If it is not possible to measure the inductor or capacitors, use the listed values.

Table 30–1

	Listed Value	Measured Value
L_1	100 mH	
C_1	0.01 μF	
C_2	1000 pF	
R_1	10 kΩ	

2. Construct the circuit shown in Figure 30–3. The 10 kΩ resistor is large compared to the Thevenin impedance of the generator. Set the generator for a 1.0 V_{pp} square wave with no load at a frequency of 1.0 kHz. You should observe that the capacitor fully charges and discharges at this frequency because the *RC* time constant is short compared to the period. On Plot 30–1, sketch the waveforms for the generator, the capacitor, and the resistor. Label voltage and time on your sketch. To look at the voltage across the resistor, use the difference function technique described in Experiment 16. The scope should be dc coupled for those measurements.

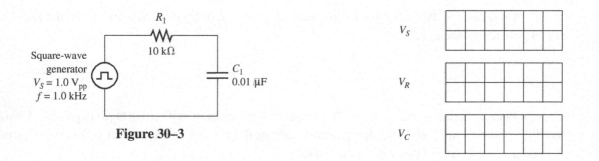

Figure 30–3

Plot 30–1

3. Compute the *RC* time constant for the circuit. Include the generator's Thevenin impedance as part of the resistance in the computation. Enter the computed time constant in Table 30–2.

Table 30–2

	Computed	Measured
RC time constant		

4. Measure the *RC* time constant using the following procedure:
   (a) With the generator disconnected from the circuit, set the output square wave on the oscilloscope to cover 5 vertical divisions (0 to 100%).
   (b) Connect the generator to the circuit. Adjust the SEC/DIV and trigger controls to stretch the capacitor-charging waveform across the scope face to obtain best resolution.
   (c) Count the number of horizontal divisions from the start of the rise to the point where the waveform crosses 3.15 *vertical* divisions (63% of the final level). Multiply the number of *horizontal* divisions that you counted by the setting of the SEC/DIV control. Alternatively, if you have cursor measurements on your oscilloscope, you may find they allow you to make a more precise measurement. Enter the measured *RC* time constant in Table 30–2.

5. Observe the capacitor waveform while you increase the generator frequency to 10 kHz. On Plot 30–2, sketch the waveforms for the generator, the capacitor, and the resistor at 10 kHz. Label the voltage and time on your sketch.

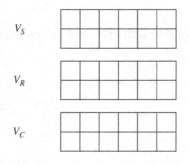

Plot 30–2

6. Temporarily, change the generator from a square wave to a triangle waveform. Describe the waveform across the capacitor.

7. Change back to a square wave at 10 kHz. Replace the capacitor with a 1000 pF capacitor. Using the difference channel, observe the waveform across the resistor. If the output were taken across the resistor, what would this circuit be called? _____

8. Replace the 1000 pF capacitor with a 100 mH inductor. Using the 10 kHz square wave, look at the signal across the generator, the inductor, and the resistor. On Plot 30–3, sketch the waveforms for each. Label the voltage and time on your sketch.

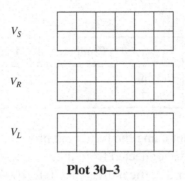

Plot 30–3

CONCLUSION

EVALUATION AND REVIEW QUESTIONS

1. (a) Explain why the Thevenin impedance of the generator was included in the calculated RC time constant measurement in step 3.

    (b) Suggest how you might find the value of an unknown capacitor using the RC time constant.

2. (a) Compute the percent difference between the measured and computed *RC* time constant.

   (b) List some factors that affect the accuracy of the measured result.

3. What accounts for the change in the capacitor voltage waveform as the frequency was raised in step 5?

4. (a) Draw an *RC* integrating circuit and an *RC* differentiating circuit.

   (b) Draw an *RL* integrating circuit and an *RL* differentiating circuit.

5. Assume you had connected a square wave to an oscilloscope but saw a signal that was differentiated as illustrated in Figure 30–2. What could account for this effect?

FOR FURTHER INVESTIGATION

The rate at which a capacitor charges is determined by the RC time constant of the equivalent series resistance and capacitance. The RC time constant for the circuit illustrated in Figure 30–4 can be determined by applying Thevenin's theorem to the left of points **A–A.** The Thevenin resistance of the generator is part of the charging path and *should* be included. The capacitor is not charging to the generator voltage but to a voltage determined by the voltage divider consisting of R_1 and R_2. Predict the time constant and the waveforms across each resistor. Investigate carefully the waveforms across each of the components.

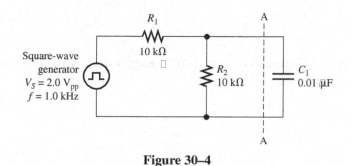

Figure 30–4

MULTISIM TROUBLESHOOTING

This experiment has four Multisim files on the website (www.prenhall.com/floyd). Three of the four files contain a simulated "fault"; one has "no fault". The file with no fault is named EXP30-3-nf. You may want to open this file to compare your results with the computer simulation. Then open each of the files with faults. Use the simulated instruments to investigate the circuit and determine the problem. The following are the filenames for circuits with troubleshooting problems for this experiment.

EXP30-3-f1

    Fault: _____

EXP30-3-f2

    Fault: _____

EXP30-3-f3

    Fault: _____

Application Assignment 15

Name _____

Date _____

Class _____

REFERENCE

Text, Chapter 15; Application Assignment: Putting Your Knowledge to Work

Step 1   From the list of standard capacitors, specify the five capacitors for the integrator delay circuit.

$C_1 = $ _____ $C_2 = $ _____ $C_3 = $ _____

$C_4 = $ _____ $C_5 = $ _____

Step 2   Complete the wire list for the breadboard using the circled numbers. The first one has been completed as an example.

From	To	From	To	From	To
1	5				

Step 3   Specify the amplitude, frequency, and duty cycle settings for the function generator in order to test the delay times.

Amplitude = _____

Frequencies for each delay time:

$f_1 = $ _____ $f_2 = $ _____

$f_3 = $ _____ $f_4 = $ _____ $f_5 = $ _____

Develop a test procedure.

Step 4   Explain how you will verify that each switch setting produces the proper output delay time.

RELATED EXPERIMENT

MATERIALS NEEDED
One 7414 hex inverter (Schmitt trigger)
One 10 kΩ potentiometer
One 0.01 μF capacitor

DISCUSSION
An interesting variation of the application assignment uses a Schmitt trigger as a switching device. A Schmitt trigger is a circuit with two thresholds for change. The switching level is dependent on whether the input signal is rising or falling. Consider the circuit shown in Figure AA–15–1. The charging and discharging of the capacitor is determined by the switching points of the Schmitt trigger. The input voltage is initially low and the output voltage is high (near 5.0 V). The capacitor begins to charge toward the higher output voltage. As the capacitor charges, the input voltage passes a trip point, causing the input voltage to go high and the output voltage to go low. The capacitor begins to discharge toward the lower voltage until it passes the lower trip point causing the process to repeat.

Construct the circuit and measure the output waveform and the waveform across the capacitor. Try varying R as you observe the capacitor voltage. What is the output waveshape and frequency? Can you determine the threshold voltages of the Schmitt trigger by observing the output?

EXPERIMENTAL RESULTS

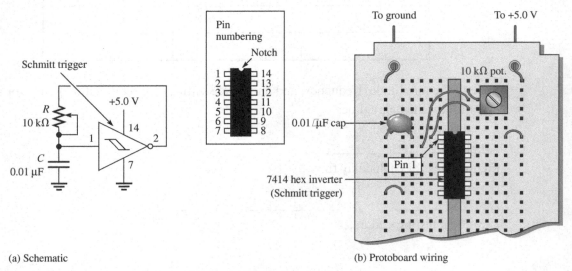

(a) Schematic

(b) Protoboard wiring

Figure AA–15–1

Checkup 15

REFERENCE
Text, Chapter 15; Lab manual, Experiment 30

1. A circuit that can be used to change a square wave into a triangle wave is:
   (a) a tuned circuit   (b) a ladder circuit   (c) an integrator   (d) a differentiator

2. For an *RC* circuit driven by a pulse, the capacitor will fully charge in:
   (a) one time constant   (b) 1 s
   (c) five time constants   (d) a time depending on the amplitude of the pulse

3. An *RC* integrator circuit is driven by a square wave that goes from 0 V to 10 V. The time constant is very short compared to the input square wave. The output will be:
   (a) a 5 V dc level   (b) a triangle waveform
   (c) an exponentially rising and falling waveform   (d) a series of positive and negative spikes

4. Assume an *RC* differentiator circuit is driven by a square wave. The output is a square wave with a slight droop and overshoot. The time constant of the circuit (τ) is:
   (a) much longer than the pulse width   (b) much shorter than the pulse width
   (c) equal to the pulse width

5. A 4.7 kΩ resistor is connected in series with a 0.1 μF capacitor. The time constant is:
   (a) 0.47 μs   (b) 0.47 ms   (c) 21.3 ns   (d) 21.3 μs

6. A 4.7 kΩ resistor is connected in series with a 10 mH inductor. The time constant is:
   (a) 0.47 μs   (b) 47 ms   (c) 4.7 s   (d) 2.13 μs

7. Assume that a switch is closed in a series *RC* circuit that has a time constant of 10 ms. The current in the circuit will be 37% of its initial value in:
   (a) 1.0 ms   (b) 3.7 ms   (c) 6.3 ms   (d) 10 ms

8. When a single pulse is applied to a series *RL* circuit, the greatest *change* in current occurs:
   (a) at the beginning   (b) at the 50% point
   (c) after one time constant   (d) at the end

9. A 1.0 kHz square wave is applied to an *RL* differentiator circuit. The current in the circuit will reach steady-state conditions if the time constant is equal to:
   (a) 100 μs   (b) 1.0 ms   (c) 6.3 ms   (d) 10 ms

10.   If you need to couple a square wave into a circuit through a capacitor, for best fidelity you should have an *RC* time constant that is:
(a) very short
(b) equal to the rise time of the square wave
(c) equal to the pulse width of the square wave
(d) very long

11.   Explain how you could use a known resistor, a square wave generator, and an oscilloscope to find the value of an unknown inductor.

12.   Assume you wanted to lengthen the time constant for the circuit in Figure 30–3 to 330 μs.
(a)   What change would you make to the circuit?

(b)   Draw an *RL* integrator with a 10 kΩ resistor and a 330 μs time constant.

13.   (a)   Compute the time constant for the circuit shown in Figure C–15–1.

(b)   What is the maximum frequency for the pulse generator in order to allow the capacitor time to charge and discharge fully?

(c)   Assume the generator is set to the frequency determined in (b); sketch the waveform across the capacitor. Show the voltage and time on your sketch.

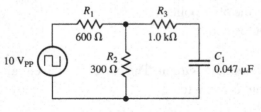

Figure C–15–1

31 Diode Characteristics

READING

Text, Sections 16–1 through 16–3

NOTE: Text references in Experiments 31–44 are only for *Electronics Fundamentals: Circuits, Devices, and Applications*.

OBJECTIVES

After performing this experiment, you will be able to:
1. Measure and plot the forward- and reverse-biased *IV* characteristics for a diode.
2. Test the effect of heat on a diode's response.
3. Measure the ac resistance of a diode.

MATERIALS NEEDED

Resistors:

One 330 Ω; one 1.0 MΩ

One signal diode (1N914 or equivalent)

SUMMARY OF THEORY

Semiconductors are certain crystalline materials that can be altered with impurities to radically change their electrical characteristics. The impurity can be an electron donor or an electron acceptor. Donor impurities provide an *extra* electron that is free to move through the crystal at normal temperatures. The total crystal is electrically neutral, but the availability of free electrons in the material causes the material to be classified as an N-type (for negative) semiconductor. Acceptor impurities leave a "hole" (the absence of an electron) in the crystal structure. These materials are called P-type (for positive) semiconductors. They conduct by the motion of shared valence bond electrons moving between the atoms of the crystal. This motion is referred to as hole motion because the absence of an electron from the crystal structure can be thought of as a hole.

When a P-type and an N-type material are effectively made on the same crystal base, a *diode* is formed. The PN junction has unique electrical characteristics. Electrons and holes diffuse across the junction, creating a *barrier potential,* which prevents further current without an external voltage source. If a dc voltage source is connected to the diode, the direction it is connected has the effect of either increasing or decreasing the barrier potential. The effect is to allow the diode to either conduct readily or to become a poor conductor. If the negative terminal of the source is connected to the N-type material and the positive terminal is connected to P-type material, the diode is said to be forward-biased, and it conducts. If the positive terminal of the source is connected to the N-type material and the negative terminal is connected to P-type material, the diode is said to be reverse-biased, and the diode is a poor conductor.

While the actual processes that occur in a diode are rather complex, diode operation can be simplified with three approximations. The first approximation is to consider the diode as a switch. If it is forward-biased, the switch is closed. If it is reverse-biased, the switch is open. The second approximation is the same as the first except it takes into account the barrier potential. For a silicon diode, this is approximately 0.7 V. A forward-biased silicon diode will drop approximately 0.7 V across the diode. The third approximation includes the first and second approximations and adds the small forward *(bulk)* resistance that is present when the diode is forward-biased.

PROCEDURE

1. Measure and record the resistance of the resistors listed in Table 31–1. Then check your diode with the ohmmeter. Select a low ohm range and measure the forward and reverse resistance by reversing the diode. The diode is good on this test if the resistance is significantly different between the forward and the reverse directions. If you are using an autoranging meter, the meter may not produce enough voltage to overcome the barrier potential. You should select a low ohm range and hold that range. Consult the operator's manual for specific instructions. Record the data in Table 31–1.

Table 31–1

Component	Listed Value	Measured Value
R_1	330 Ω	
R_2	1.0 MΩ	
D_1 forward resistance		
D_1 reverse resistance		

2. Construct the forward-biased circuit shown in Figure 31–1. The line on the diode indicates the cathode side of the diode. Set the power supply for zero volts.

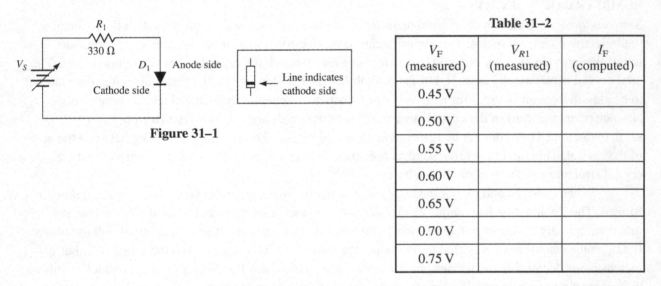

Figure 31–1

Table 31–2

V_F (measured)	V_{R1} (measured)	I_F (computed)
0.45 V		
0.50 V		
0.55 V		
0.60 V		
0.65 V		
0.70 V		
0.75 V		

3. Monitor the forward voltage drop, V_F, *across the diode*. Slowly increase V_S to establish 0.45 V across the diode. Measure the voltage across the resistor, V_{R1}, and record it in Table 31–2.

4.   The diode forward current, I_F, can be found by applying Ohm's law to R_1. Compute I_F and enter the computed current in Table 31–2.

5.   Repeat steps 3 and 4 for each voltage listed in Table 31–2.

6.   With the power supply set to the voltage that causes 0.75 V to drop across the diode, bring a hot soldering iron near the diode. Do *not* touch the diode with the iron. Observe the effect of heat on the voltage and current in a forward-biased diode. If you have freeze spray available, test the effect of freeze spray on the diode's operation. Describe your observations.

7.   The data in this step will be accurate only if your voltmeter has a very high input impedance. You can find out if your meter is high impedance by measuring the power supply voltage through a series 1.0 MΩ resistor. If the meter reads the supply voltage accurately, it has high input impedance. Connect the reverse-biased circuit shown in Figure 31–2. Set the power supply to each reverse voltage listed in Table 31–3, (V_R). Measure and record the voltage across R_2 (V_{R2}). Use this voltage and Ohm's law to compute the reverse current in each case. Enter the computed current in Table 31–3.

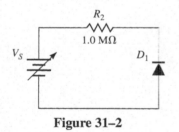

Figure 31–2

Table 31–3

V_R (measured)	V_{R2} (measured)	I_R (computed)
5.0 V		
10.0 V		
15.0 V		

8.	Graph the forward- and reverse-biased diode curves on Plot 31–1. The different voltage scale factors for the forward and reverse curves are chosen to allow the data to cover more of the graph. You need to choose an appropriate current scale factor that will put the largest current recorded near the top of the graph.

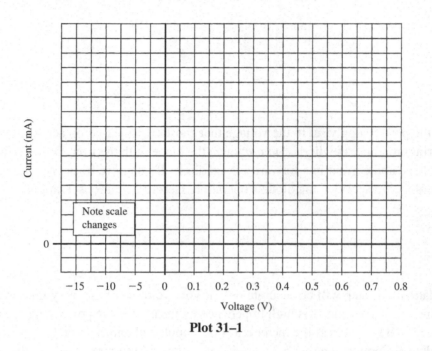

Plot 31–1

9.	With the power supply set to 15 V, bring a hot soldering iron near the diode. Do *not* touch the diode with the iron. Observe the effect of heat on the voltage and current in the reverse-biased diode. If you have freeze spray available, test the effect of freeze spray on the diode's operation. Describe your observations.

CONCLUSION

EVALUATION AND REVIEW QUESTIONS

1.   Compute the diode's forward resistance at three points on the forward-biased curve. Apply Ohm's law to the curve in Plot 31–1 at 0.5 V, 0.6 V, and 0.7 V by dividing a small change in voltage by a small change in current, as illustrated in Figure 31–3. This result is called the ac resistance (r_{ac}) of the diode.

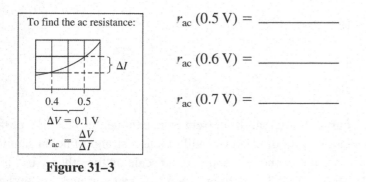

r_{ac} (0.5 V) = _____

r_{ac} (0.6 V) = _____

r_{ac} (0.7 V) = _____

Figure 31–3

2.   Does the diode's reverse resistance stay constant? Explain your answer.

3.   From the data in Table 31–2, compute the maximum power dissipated in the diode.

4.   Based on your observations of the heating and cooling of a diode, what does heat do to the forward and reverse resistance of a diode?

5.   Explain how you could use an ohmmeter to identify the cathode of an unmarked diode. Why is it necessary to know the *actual* polarity of the ohmmeter leads?

6.   A student measures the resistance of an unmarked diode with an ohmmeter. When the (+) lead of the ohmmeter is connected to lead 1 of the diode and the (−) lead of the ohmmeter is connected to lead 2 of the diode, the reading is 400 Ω. When the ohmmeter leads are reversed, the reading is ∞. Which lead on the diode is the anode?

FOR FURTHER INVESTIGATION

The theoretical equation for a diode's *I-V* curve shows that the current is an exponential function of the bias voltage.[1] This means that the theoretical forward diode curve will plot as a straight line on semilog paper. Semilog paper contains a logarithmic scale on one axis and a linear scale on the other axis. Graph your data from this experiment (Table 31–2) onto Plot 31–2. What conclusion can you make from the data you recorded?

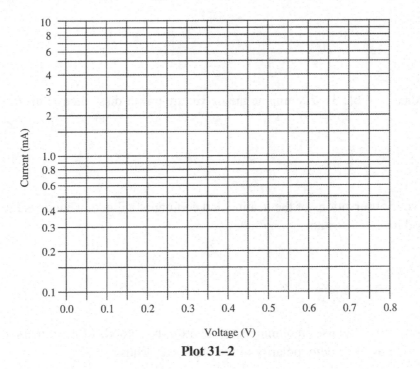

Plot 31–2

---

[1]A complete discussion of the diode equation can be found in Bogart, *Electronic Devices and Circuits,* 6th edition, 2004, Prentice Hall.

32 Rectifier Circuits

READING
Text, Sections 16–4 through 16–7

OBJECTIVES
After performing this experiment, you will be able to:
1. Construct half-wave, full-wave, and bridge rectifier circuits, and compare the input and output voltage for each.
2. Connect a filter capacitor to each circuit in objective 1 and measure the ripple voltage and ripple frequency.

MATERIALS NEEDED
One 12.6 V$_{rms}$ center-tapped transformer with fused line cord
Four diodes 1N4001 (or equivalent)
Two 2.2 kΩ resistors
One 100 μF capacitor
For Further Investigation:
   One 0.01 μF capacitor
   One 7812 or 78L12 regulator

SUMMARY OF THEORY
Rectifiers are diodes used to change ac to dc. They work like a one-way valve, allowing current in only one direction, as illustrated in Figure 32–1. The diode is forward-biased for one half-cycle of the applied voltage and reverse-biased for the other half-cycle. The output waveform is a pulsating dc wave. This waveform can then be filtered to remove the unwanted variations.

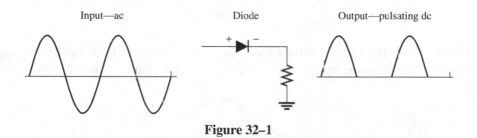

Figure 32–1

    Rectifiers are widely used in power supplies that provide the dc voltage necessary for almost all active devices to work. The three basic rectifier circuits are the half-wave, the center-tapped full-wave, and the full-wave bridge rectifier circuits. The most important parameters for choosing diodes for these circuits are the maximum forward current, I_F, and the peak inverse voltage rating (PIV) of the diode. The peak inverse voltage is the maximum voltage the diode can withstand when it is reverse-biased. The amount of reverse voltage that appears across a diode depends on the type of circuit in which it is connected. Some characteristics of the three rectifier circuits are investigated in this experiment.

Rectifier circuits are generally connected through a transformer, as shown in Figure 32–2. Notice the ground on the primary side of the transformer is not the same as the ground on the secondary side of the transformer. This is because transformers isolate the ground connection of the 3-wire service connection. The oscilloscope chassis is normally connected to earth ground through the 3-prong service cord, causing the ground to be common; however, you cannot be certain of this. If there is no connection between the grounds, the reference ground is said to be a *floating* ground. You can determine if the ground is floating by testing the voltage difference between the grounds.

PROCEDURE

Do this experiment only under supervision.

> **Caution!** In this experiment, you are instructed to connect a low-voltage (12.6 V ac) transformer to the ac line. Be certain that you are using a properly fused and grounded transformer that has no exposed primary leads. Do not touch any connection in the circuit. At no time will you make a measurement on the primary side of the transformer. Have your connections checked by your instructor before applying power to the circuit.

1. Connect the half-wave rectifier circuit shown in Figure 32–2. Notice the polarity of the diode. Connect the oscilloscope so that channel 1 is across the transformer secondary and channel 2 is across the load resistor. The oscilloscope should be set for LINE triggering since all waveforms are synchronized with the ac line voltage. View the secondary voltage, V_{SEC}, and load voltage, V_{LOAD}, for this circuit and sketch them on Plot 32–1. Label voltages on your sketch.

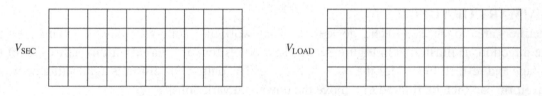

V_{SEC} V_{LOAD}

Plot 32–1

Measure the rms input voltage to the diode, V_{SEC}, and the output peak voltage, V_{LOAD}. Remember to convert the oscilloscope reading of V_{SEC} to rms. Record the data in Table 32–1.

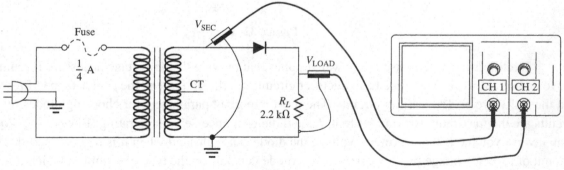

Figure 32–2

Table 32–1 Half-wave rectifier circuit.

Without Filter Capacitor				With Filter Capacitor		
Computed	Measured	Computed	Measured	Measured		Ripple Frequency
V_{IN} (rms)	V_{SEC} (rms)	V_{LOAD} (peak)	V_{LOAD} (peak)	V_{LOAD} (dc)	V_{RIPPLE}	
12.6 V ac						

2. The output isn't very useful as a dc source because of the variations in the output waveform. Connect a 100 μF capacitor (C_1) in parallel with the load resistor (R_L). *Note the polarity of the capacitor.* Measure the dc load voltage, V_{LOAD}, and the peak-to-peak ripple voltage, V_{RIPPLE}, in the output. To measure the ripple voltage, switch the oscilloscope vertical input to AC COUPLING. This allows you to magnify the small ac ripple voltage without including the much larger dc level. Measure the peak-to-peak ripple voltage and the ripple frequency. The ripple frequency is the frequency at which the waveform repeats. Record all data in Table 32–1.

3. Disconnect power and change the circuit to the full-wave rectifier circuit shown in Figure 32–3. Notice that the ground for the circuit has changed. The oscilloscope ground needs to be connected as shown. *Check your circuit carefully before applying power.* Compute the expected peak output voltage. Then apply power and view the V_{SEC} and V_{LOAD} waveforms. Sketch the observed waveforms on Plot 32–2.

V_{SEC} V_{LOAD}

Plot 32–2

Measure V_{SEC} (rms) and the peak output voltage (V_{LOAD}) without a filter capacitor. Record the data in Table 32–2.

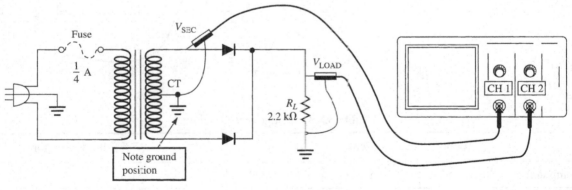

Figure 32–3

265

Table 32–2 Full-wave rectifier circuit.

Without Filter Capacitor				With Filter Capacitor		
Computed	Measured	Computed	Measured	Measured		Ripple Frequency
V_{SEC} (rms)	V_{SEC} (rms)	V_{LOAD} (peak)	V_{LOAD} (peak)	V_{LOAD} (dc)	V_{RIPPLE}	
6.3 V ac						

4.   Connect the 100 μF capacitor in parallel with the load resistor. Measure V_{LOAD}, the peak-to-peak ripple voltage, and the ripple frequency as before. Record the data in Table 32–2.

5.   Investigate the effect of the load resistor on the ripple voltage by connecting a second 2.2 kΩ load resistor in parallel with R_L and C_1 in the full-wave circuit in Figure 32–3. Measure the ripple voltage. What can you conclude about the effect of additional load current on the ripple voltage?

6.   Disconnect power and change the circuit to the bridge rectifier circuit shown in Figure 32–4. Notice that *no* terminal of the transformer secondary is at ground potential. The input voltage to the bridge, V_{SEC}, is not referenced to ground. *The oscilloscope cannot be used to view both the input voltage and the load voltage at the same time.* Check your circuit carefully before applying power. Compute the expected peak output voltage. Then apply power and *use a voltmeter* to measure V_{SEC} (rms). Use the oscilloscope to measure the peak output voltage (V_{LOAD}) without a filter capacitor. Record the data in Table 32–3.

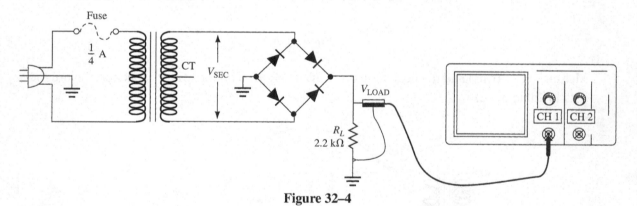

Figure 32–4

Table 32–3 Bridge rectifier circuit.

Without Filter Capacitor				With Filter Capacitor		
Computed	Measured	Computed	Measured	Measured		Ripple Frequency
V_{SEC} (rms)	V_{SEC} (rms)	V_{LOAD} (peak)	V_{LOAD} (peak)	V_{LOAD} (dc)	V_{RIPPLE}	
12.6 V ac						

7. Connect the 100 μF capacitor in parallel with the load resistor. Measure V_{LOAD}, the peak-to-peak ripple voltage, and the ripple frequency as before. Record the data in Table 32–3.

8. Simulate an open diode in the bridge by removing one diode from the circuit. What happens to the output voltage? The ripple voltage? The ripple frequency?

CONCLUSION

EVALUATION AND REVIEW QUESTIONS
1. What advantage does a full-wave rectifier circuit have over a half-wave rectifier circuit?

2. Compare a bridge rectifier circuit with a full-wave rectifier circuit. Which has the higher output voltage? Which has the greater current in the diodes?

3. Explain how you could measure the ripple frequency to determine if a diode were open in a bridge rectifier circuit.

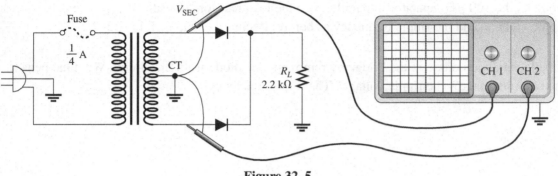

Figure 32–5

4. In step 3, you moved the ground reference to the center-tap of the transformer. If you wanted to look at the voltage across the entire secondary, you would need to connect the oscilloscope as shown in Figure 32–5 and *subtract* channel 2 from channel 1. (Some oscilloscopes do not have this capability.) Why is it necessary to use *two* channels to view the entire secondary voltage?

5. (a) What is the maximum dc voltage you could expect to obtain from a transformer with an 18 V_{rms} secondary using a bridge circuit with a filter capacitor?

    (b) What is the maximum dc voltage you could expect to obtain from the same transformer connected in a full-wave rectifier circuit with a filter capacitor?

FOR FURTHER INVESTIGATION

The bridge rectifier circuit shown in Figure 32–4 can be changed to a +12 V regulated power supply with the addition of a 7812 or 78L12 three-terminal regulator. With proper heat sinking, the 7812 can deliver over 1.0 A of current; the 78L12 can deliver over 100 mA. Add one of the regulators to your bridge rectifier circuit as shown in Figure 32–6. Measure the output ripple from the circuit with the regulator. Compare your results with the unregulated circuit in step 7.

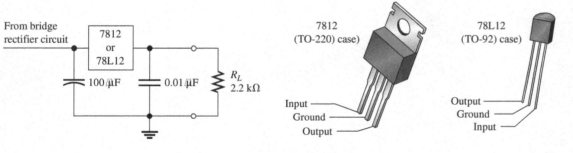

Figure 32–6

MULTISIM TROUBLESHOOTING

This experiment has four Multisim files on the website (www.prenhall.com/floyd). Three of the four files contain a simulated "fault"; one has "no fault". The file with no fault is named EXP32-3-nf. You may want to open this file to compare your results with the computer simulation. Then open each of the files with faults. Use the simulated instruments to investigate the circuit and determine the problem. The following are the filenames for circuits with troubleshooting problems for this experiment.

EXP32-3-f1

Fault: _____

EXP32-3-f2

Fault: _____

EXP32-3-f3

Fault: _____

Application Assignment 16

Name _____
Date _____
Class _____

REFERENCE

Text, Chapter 16; Application Assignment: Putting Your Knowledge to Work

Step 1 Identify the components.

Step 2 Relate the PC boards to the schematics.

Step 3 Analyze the power supply/IR emitter board. Determine the voltages that are at points 1, 2, and 3. Assume the boards are operating correctly.

$V_1 = $ _____ $V_2 = $ _____ $V_3 = $ _____

Step 4 Analyze the IR detector board. Determine the voltage at point 4 with no light. (Neglect the small dark current and current into the digital circuit.)

$V_4 = $ _____ (with no incident light)

Step 5 Determine the voltage at point 4 with incident light. Assume a reverse current of 10 μA through the photodiode.

$V_4 = $ _____ (with incident light)

Step 6 Troubleshoot the system. Give the probable cause of each fault listed in the text. Complete Table AA–16–1.

Table AA–16–1

Fault	Probable Cause
1	
2	
3	
4	
5	
6	
7	
8	

RELATED EXPERIMENT

MATERIALS NEEDED
One opto-coupler (4N35 or equivalent)
One 10 μF capacitor
Resistors:
       One 220 Ω, one 1.0 kΩ, one 2.2 kΩ

DISCUSSION
The test circuit shows an infrared LED and photodiode combination used to count objects. This combination of light source and receiver are also used when it is necessary to achieve a high degree of electrical isolation. (An electrocardiogram machine is an example.) Devices are available in packages called an opto-coupler (or opto-isolator) containing an LED light source and a sensor (a phototransistor or photodiode) in an opaque enclosure. In this application assignment, an audio signal will be transmitted by light in an optical isolator to a phototransistor connected as a photodiode. Normally, an opto-coupler is used to isolate two circuits, but in order to simplify the test circuit, common grounds and supply voltages are used here. Set up the test circuit shown in Figure AA–16–1. Set the function generator for a 1.0 V_{pp} sine wave at 1.0 kHz. This causes the LED to vary in brightness at a rate determined by the function generator. Compare the input and output signals on a two-channel oscilloscope. Compare the phase of the input and output signal and note the amplitude of the received signal. Summarize your observations.

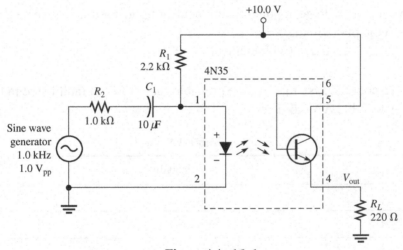

Figure AA–16–1

Checkup 16

REFERENCE
Text, Chapter 16; Lab manual, Experiments 31 and 32

1. A P-type material is a semiconductor containing:
    (a) only pure material   (b) extra electrons
    (c) donor impurities   (d) acceptor impurities

2. The number of valence electrons in the outer shell of a donor impurity is:
    (a) 1   (b) 3   (c) 4   (d) 5

3. An example of an acceptor impurity is:
    (a) aluminum   (b) germanium   (c) arsenic   (d) phosphorus

4. The ac resistance of a silicon diode is highest:
    (a) in the reverse direction   (b) at 0.5 V
    (c) at 0.6 V   (d) at 0.7 V

5. When a forward-bias voltage is applied to a PN junction, the barrier voltage is
    (a) reduced   (b) unchanged   (c) increased

6. An advantage of a full-wave bridge rectifier is:
    (a) it uses four diodes   (b) two diodes are in series with the load
    (c) diodes carry a smaller current   (d) a center-tapped transformer is not needed

7. A half-wave rectified sine wave has an average voltage of 25 V. The peak voltage is
    approximately:
    (a) 8 V   (b) 16 V   (c) 39 V   (d) 79 V

8. A full-wave rectifier supply contains a capacitor-input filter. The amount of the ripple voltage is
    dependent on the size of:
    (a) the capacitor   (b) the load resistor
    (c) the capacitor and load resistor   (d) neither the capacitor nor load resistor

9. If one diode in a full-wave rectifier is open, the output:
    (a) ripple voltage and frequency will both increase
    (b) ripple voltage will increase, but frequency will decrease
    (c) ripple voltage and frequency will both decrease
    (d) ripple voltage will decrease, but frequency will increase

10. A full-wave bridge rectifier is across the 12.6 V_{rms} secondary of a transformer. The output is
    filtered with a capacitor-input filter. The dc voltage is about:
    (a) 9 V   (b) 12 V   (c) 17 V   (d) 25 V

11. If you check the forward resistance of a diode with an ohmmeter, it will change depending on the range setting of the meter. Explain.

12. For the circuit shown in Figure C–16–1, determine if there is a fault for each of the following conditions. If so, state what the most likely problem is.
    (a)   The voltmeter reads zero; the ammeter reads 3.0 mA.

    (b)   The voltmeter reads +10 V; the ammeter reads zero.

    (c)   The voltmeter reads 0.7 V; the ammeter reads 2.8 mA.

    (d)   Both the voltmeter and the ammeter read zero.

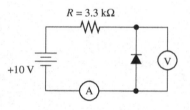

Figure C–16–1

13. (a) Explain how a measurement of the ripple frequency can help determine if all diodes are working in a bridge circuit.

    (b) Assume that you discovered from the ripple measurement that a diode was open, how would you locate which diode was bad?

14. A transformer with a 12.6 V_{rms} secondary supplies the ac voltage for a full-wave bridge circuit. The output is connected to a 1000 µF capacitor in parallel with a 4.7 kΩ resistive load.
    (a)   Compute the dc voltage on the load. _____

    (b)   Compute the dc load current. _____

274

33 Bipolar Junction Transistors

Name _____

Date _____

Class _____

READING
Text, Section 17–1 and first part of Section 17–2 (to load line operation)

OBJECTIVES
After performing this experiment, you will be able to:
1.   Measure and graph the collector characteristic curves for a bipolar junction transistor.
2.   Use the characteristic curves to determine the β_{dc} of the transistor at a given point.

MATERIALS NEEDED
Resistors:
        One 100 Ω, one 33 kΩ
One 2N3904 NPN transistor (or equivalent)
For Further Investigation:
    Transistor curve tracer

SUMMARY OF THEORY
A bipolar junction transistor (BJT) is a three-terminal device capable of amplifying an ac signal. The three terminals are called the base, emitter, and the collector. BJTs consist of a very thin base material sandwiched in between two of the opposite type materials. They are available in two forms, either NPN or PNP. The middle letter indicates the type of material used for the base, while the outer letters indicate the emitter and collector material. Two PN junctions are formed when a transistor is made, the junction between the base and emitter and the junction between the base and collector. These two junctions form two diodes, the emitter-base diode and the base-collector diode.

BJTs are current amplifiers. A small base current is amplified to a larger current in the collector-emitter circuit. An important characteristic is the dc current gain, which is the ratio of collector current to base current. This is called the dc beta (β_{dc}) of the transistor. Another useful characteristic is the dc alpha (α_{dc}). The dc alpha is the ratio of the collector current to the emitter current and is always less than 1.

For a transistor to amplify, power is required from dc sources. The dc voltages required for proper operation are referred to as bias voltages. The purpose of bias is to establish and maintain the required operating conditions despite variations between transistors or circuit parameters. For normal operation, the base-emitter junction is forward-biased and the base-collector junction is reverse-biased. Since the base-emitter junction is forward-biased, it has characteristics of a forward-biased diode. A silicon bipolar transistor requires approximately 0.7 V of voltage across the base-emitter junction for there to be base current.

PROCEDURE

1. Measure and record the resistance of the resistors listed in Table 33–1.

Table 33–1

	Listed Value	Measured Value
R_1	33 kΩ	
R_2	100 Ω	

2. Connect the common emitter configuration illustrated in Figure 33–1. Start with both power supplies set to 0 V. The purpose of R_1 is to limit base current and allow determination of the base current. Slowly increase V_{BB} until V_{R1} is 1.65 V. This sets up a base current of 50 μA, which can be shown by applying Ohm's law to R_1.

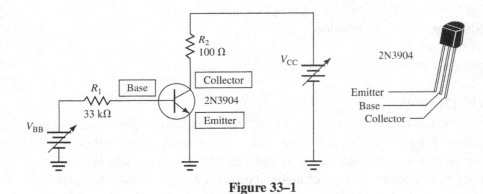

Figure 33–1

3. Without disturbing the setting of V_{BB}, slowly increase V_{CC} until 2.0 V is measured between the transistor's collector and emitter. This voltage is called V_{CE}. Then measure and record V_{R2} for this setting. Record V_{R2} in Table 33–2 under the columns labeled Base Current = 50 μA.

Table 33–2

	Base Current = 50 μA		Base Current = 100 μA		Base Current = 150 μA	
V_{CE} (measured)	V_{R2} (measured)	I_C (computed)	V_{R2} (measured)	I_C (computed)	V_{R2} (measured)	I_C (computed)
2.0 V						
4.0 V						
6.0 V						
8.0 V						

4. Compute the collector current, I_C, by applying Ohm's law to R_2. Use the measured voltage, V_{R2}, and the measured resistance, R_2, to determine the current. Note that the current in R_2 is the same as I_C for the transistor. Enter the computed collector current in Table 33–2 under the columns labeled Base Current = 50 μA.

5. Without disturbing the setting of V_{BB}, increase V_{CC} until 4.0 V is measured across the transistor's collector to emitter. Measure and record V_{R2} for this setting. Compute the collector current by applying Ohm's law as in step 4. Continue in this manner for each of the values of V_{CE} listed in Table 33–2.

6. Reset V_{CC} for 0 V and adjust V_{BB} until V_{R1} is 3.3 V. The base current is now 100 μA.

7. Without disturbing the setting of V_{BB}, slowly increase V_{CC} until V_{CE} is 2.0 V. Then measure and record V_{R2} for this setting in Table 33–2 under columns labeled Base Current = 100 μA. Compute I_C for this setting by applying Ohm's law to R_2. Enter the computed collector current in Table 33–2.

8. Increase V_{CC} until V_{CE} is equal to 4.0 V. Measure and record V_{R2} for this setting. Compute I_C as before. Continue in this manner for each value of V_{CE} listed in Table 33–2.

9. Reset V_{CC} for 0 V and adjust V_{BB} until V_{R1} is 4.95 V. The base current is now 150 μA.

10. Complete Table 33–2 by repeating steps 7 and 8 for 150 μA of base current.

11. Plot three collector characteristic curves using the data tabulated in Table 33–2. The collector characteristic curve is a graph of V_{CE} versus I_C for a constant base current. Choose a scale for I_C that allows the largest current observed to fit on the graph. Label each curve with the base current it represents. Graph the data on Plot 33–1.

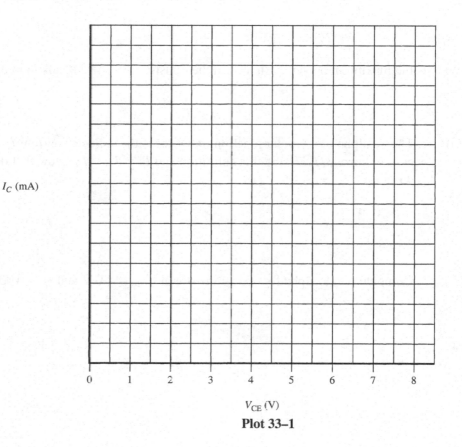

V_{CE} (V)

Plot 33–1

277

12. Use the characteristic curve you plotted to determine the β_{dc} for the transistor at a V_{CE} of 3.0 V and a base current of 50 µA, 100 µA, and 150 µA. Then repeat the procedure for a β_{dc} at a V_{CE} of 5.0 V. Tabulate your results in Table 33–3.

Table 33–3

V_{CE}	Current Gain, β_{dc}		
	$I_B = 50$ µA	$I_B = 100$ µA	$I_B = 150$ µA
3.0 V			
5.0 V			

CONCLUSION

EVALUATION AND REVIEW QUESTIONS

1. Do the experimental data indicate that β_{dc} is a constant at all points? Does this have any effect on the linearity of the transistor?

2. What effect would a higher β_{dc} have on the characteristic curves you measured?

3. What is the maximum power dissipated in the transistor for the data taken in the experiment?

4. (a) The dc alpha of a bipolar junction transistor is the collector current I_C, divided by the emitter current, I_E. Using this definition and $I_E = I_C + I_B$, show that dc alpha can be written

$$\alpha_{dc} = \frac{\beta_{dc}}{\beta_{dc} + 1}$$

   (b) Compute the dc alpha for your transistor at $V_{CE} = 4.0$ V and $I_B = 100$ µA.

5.   What value of V_{CE} would you expect if the base terminal of a transistor were open? Explain your answer.

FOR FURTHER INVESTIGATION

If you have a transistor curve tracer available, use it to check the data taken in this experiment. A transistor curve tracer has a step generator that generates a staircase set of current or voltage steps. Set the step generator to 50 μA per step. Select positive steps to apply to the base with the emitter grounded. Select a positive sweep voltage of approximately +20 V with a series limiting resistance of several hundred ohms. Select a horizontal display of 1 V/div and a vertical display of about 10 mA/div. (If your transistor has a very high or low β_{dc}, you may need to change these settings.) The curve tracer will show the collector characteristic curves. Test the effect of heating or cooling the transistor on β_{dc}.

READING
Text, Section 17–2

OBJECTIVES
After performing this experiment, you will be able to:
1. Compute the dc parameters, r_e, and the voltage gain of a common-emitter amplifier with voltage divider bias.
2. Build a common-emitter amplifier and measure the dc and ac parameters.
3. Predict the result of faults in a common-emitter amplifier.

MATERIALS NEEDED
Resistors:

One 100 Ω, one 2.2 kΩ, one 6.8 kΩ, three 10 kΩ, one 47 kΩ

Capacitors:

Two 1 μF, one 100 μF

One 10 kΩ potentiometer
One 2N3904 NPN transistor (or equivalent)

SUMMARY OF THEORY
In a common-emitter (CE) amplifier, the input signal is applied between the base and emitter, and the output signal is developed between the collector and emitter. The transistor's *emitter* is common to both the input and output circuit; hence, the term common emitter. Do not confuse the term *common emitter* with grounded emitter. The emitter terminal of a CE amplifier may or may not be at circuit ground.

To make any transistor circuit amplify ac signals, the base-emitter junction must be forward-biased, and the base-collector junction must be reverse-biased. The purpose of bias circuits is to establish and maintain the proper dc operating conditions for the transistor. The bias circuit must provide these conditions for wide variations between transistors that may occur as a result of mass production.

There are several ways to apply dc bias. The simplest method, called base bias or fixed bias, is frequently unsatisfactory due to manufacturing variations between transistors and sensitivity to temperature changes. Base bias is recognized by a single resistor connected from V_{CC} to the transistor base. A much more widely used bias circuit is called voltage-divider bias. Voltage-divider bias is not as sensitive to transistor variations and temperature changes. Voltage-divider bias is shown in Figure 34–1(a).

There are many variations in transistor amplifiers. The purpose of the example shown is to develop a *method for analysis* rather than a set of equations. The equations for each configuration of amplifier are necessarily different, depending on the circuit. You should not attempt to memorize a set of equations for analysis, but rather observe the application of analysis methods.

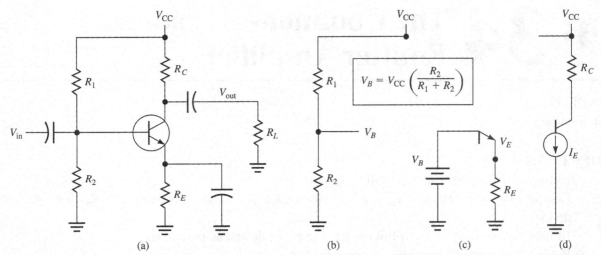

Figure 34-1 Steps in solving the dc parameters in CE amplifier with voltage-divider bias. Note that the equation given in (b) assumes a "stiff" divider—that is, it ignores the small loading effect of the base current on the divider. As long as the base current is small compared to the divider current, this is satisfactory.

To analyze any amplifier, start with the dc parameters. The steps to solve for the dc parameters for the CE amplifier with voltage-divider bias illustrated in Figure 34–1(a) are:

1. Mentally remove capacitors from the circuit since they appear open to dc. This causes the load resistor, R_L, to be removed. Solve for the base voltage, V_B, by applying the voltage divider rule to R_1 and R_2, as illustrated in Figure 34–1(b).

2. Subtract the 0.7 V forward-bias drop across the base-emitter diode from V_B to obtain the emitter voltage, V_E, as illustrated in Figure 34–1(c).

3. The dc current in the emitter circuit is found by applying Ohm's law to R_E. The emitter current, I_E, is approximately equal to the collector current, I_C. The transistor appears to be a current source of approximately I_E into the collector circuit, as shown in Figure 34–1(d).

The ac parameters for the amplifier can now be analyzed. The circuit and the ac equivalent circuit are shown in Figure 34–2. The capacitors appear to be an ac short. For this reason, the ac equivalent circuit does not contain R_E. Using the superposition theorem, V_{CC} is replaced with a short, placing it at ac ground. The analysis steps are:

1. Replace all capacitors with a short and place V_{CC} at ac ground. Compute the ac resistance of the emitter, r_e, from the equation:

$$r_e = \frac{25 \text{ mV}}{I_E}$$

2. Compute the amplifier's voltage gain. Voltage gain is the ratio of the output voltage divided by the input voltage. The input voltage is across the ac emitter resistance to ground which, in this case, is r_e. The output voltage is taken across the ac resistance from collector to ground. Looking from the transistor's collector, R_L appears to be in parallel with R_C. For the circuit in Figure 34–2(b), the output voltage divided by the input voltage can be written:

$$A_v = \frac{V_{out}}{V_{in}} = \frac{I_c(R_C \parallel R_L)}{I_e r_e} \cong \frac{R_C \parallel R_L}{r_e}$$

3.   Compute the total input resistance seen by the ac signal:

$$R_{in(T)} = R_1 \parallel R_2 \parallel \beta_{ac} r_e$$

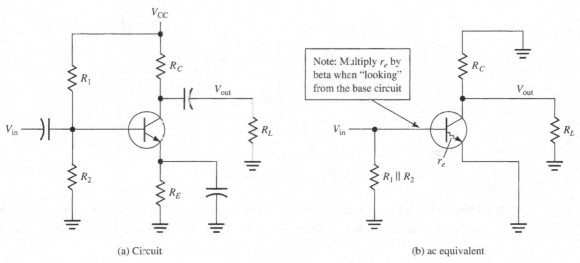

(a) Circuit   (b) ac equivalent

Figure 34–2

PROCEDURE

1.   Measure and record the resistance of the resistors listed in Table 34–1.

<table>
<tr><td colspan="3" align="center">**Table 34–1**</td></tr>
<tr><td>Resistor</td><td>Listed
Value</td><td>Measured
Value</td></tr>
<tr><td>R_1</td><td>47 kΩ</td><td></td></tr>
<tr><td>R_2</td><td>10 kΩ</td><td></td></tr>
<tr><td>R_3</td><td>10 kΩ</td><td></td></tr>
<tr><td>R_4</td><td>100 Ω</td><td></td></tr>
<tr><td>R_E</td><td>2.2 kΩ</td><td></td></tr>
<tr><td>R_C</td><td>6.8 kΩ</td><td></td></tr>
<tr><td>R_L</td><td>10 kΩ</td><td></td></tr>
</table>

<table>
<tr><td colspan="3" align="center">**Table 34–2**</td></tr>
<tr><td>DC
Parameter</td><td>Computed
Value</td><td>Measured
Value</td></tr>
<tr><td>V_B</td><td></td><td></td></tr>
<tr><td>V_E</td><td></td><td></td></tr>
<tr><td>I_E</td><td></td><td></td></tr>
<tr><td>V_C</td><td></td><td></td></tr>
<tr><td>V_{CE}</td><td></td><td></td></tr>
</table>

2.   Using measured resistances, compute the dc parameters listed in Table 34–2 for the CE amplifier shown in Figure 34–3. This circuit, like most voltage-divider bias circuits, uses a "stiff" divider; therefore, the equation shown in the box in Figure 34–1 is satisfactory for finding the base voltage. Compute the base voltage, V_B, emitter voltage, V_E, and emitter current, I_E, as described in the Summary of Theory. The emitter current is assumed to be the same as the collector current. Use this idea and Ohm's law to find the voltage drop across the collector resistor. V_C can then be found by subtracting this voltage drop from V_{CC}. V_{CE} is the difference between V_C and V_E.

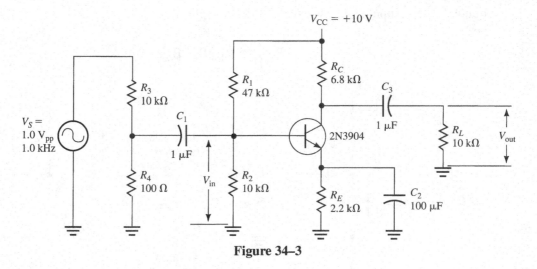

Figure 34–3

3. Construct the amplifier shown in Figure 34–3. The signal generator should be turned off. Measure and record the dc voltages listed in Table 34–2. Your measured and computed values should agree within 10%.

4. Compute the ac parameters listed in Table 34–3 using the method given in the Summary of Theory. The ac base voltage, V_b, represents the signal input to the amplifier, V_{in}. It is listed as 10 mV$_{pp}$ based on the input voltage divider consisting of R_3 and R_4. Multiply the input signal by the computed voltage gain to obtain the output signal. The ac collector voltage, V_c, represents the output signal, V_{out}. If you do not know β_{ac} for the input resistance calculation, assume a value of 100.

5. Turn on the signal generator and adjust V_S for a 1.0 V$_{pp}$ signal at 1.0 kHz. The ac input at the base, V_b, is already listed in Table 34–3 as 10 mV$_{pp}$. Measure the ac collector signal, V_c, and record it in Table 34–3. Use the ac base voltage and measured collector voltage to obtain the measured voltage gain. Record the measured voltage gain in Table 34–3.

Table 34–3

AC Parameter	Computed Value	Measured Value
$V_b = V_{in}$	10 mV$_{pp}$	10 mV*$_{pp}$
r_e		
A_v		
$V_c = V_{out}$		
$R_{in(T)}$		

*Based on setting V_S to 1.0 V$_{pp}$

6. The measurement of $R_{in(T)}$ must be done indirectly since it represents an ac resistance. The output signal (V_{out}) is monitored and noted. A variable test resistor (R_{test}) is then inserted in series with the source, as shown in Figure 34–4. The resistance of R_{test} is increased until V_{out} drops to one-half the value noted prior to inserting R_{test}. This means the voltage drop across R_{test} is equal to the voltage drop across $R_{in(T)}$; hence, the resistances are equal. R_{test} can then be removed and measured with an ohmmeter. Using this method, measure $R_{in(T)}$ and record the result in Table 34–3.

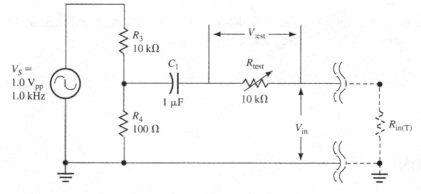

Figure 34–4   Measurement of $R_{in(T)}$

7. Restore the circuit to that of Figure 34–3. With a two-channel oscilloscope, compare the input and output waveforms. What is the phase relationship between V_{in} and V_{out}?

8. Remove C_2 from the circuit. Measure the ac signal voltage at the transistor's base, emitter, and collector. Measure the voltage gain of the amplifier. What conclusion can you make about the amplifier's performance with C_2 open?

9. Replace C_2 and remove R_L. Again measure the ac signal voltage at the transistor's base, emitter, and collector. Measure the voltage gain of the amplifier. What conclusion can you make about the amplifier's performance with R_L open?

10. Replace R_L and open R_E. Measure the dc voltages at the base, emitter, and collector. Is the transistor cut off or saturated? (Saturation is *maximum* current in the transistor; cutoff is *no* current in the transistor.) Explain your answer.

11. Replace R_E and open R_2. Measure the dc voltages at the base, emitter, and collector. Is the transistor cut off or saturated? Explain your answer.

CONCLUSION

EVALUATION AND REVIEW QUESTIONS

1. When C_2 is open, you found that the gain is affected. Explain.

2. In step 6, you were instructed to measure the input resistance while monitoring the output voltage. Why is the procedure better than monitoring the base voltage?

3. Assume the amplifier shown in Figure 34–3 has $+1.8$ V dc measured on the base, 1.1 V dc measured on the emitter, and $+1.1$ V dc measured on the collector.
   (a) Is this normal?

   (b) If not, what is the most likely cause of the problem?

4. If C_2 were shorted:
   (a) What dc base voltage would you expect?

   (b) What dc collector voltage would you expect?

5. Explain a simple test to determine if a transistor is saturated or cut off.

6. What is meant by a "stiff" voltage divider?

FOR FURTHER INVESTIGATION

The low frequency response of the CE amplifier in this experiment is determined by the coupling and bypass capacitors. The upper frequency response is determined by the unseen interelectrode and stray circuit capacitances. Using the oscilloscope to view the output waveform, set the generator for a midband frequency of 1 kHz. Use a sine wave with a convenient level (not clipped) across the load resistor. Raise the generator frequency until the output voltage falls to 70.7% of the midband level. This is the upper cutoff frequency. Then lower the generator frequency until the output voltage falls to 70.7% of the midband level. This is the lower cutoff frequency. Try switching C_1 and C_2. What effect does this have on the lower cutoff frequency? Does it have an effect on the upper cutoff frequency? Summarize your investigation in a short report.

MULTISIM TROUBLESHOOTING

This experiment has four Multisim files on the website (www.prenhall.com/floyd). Three of the four files contain a simulated "fault"; one has "no fault". The file with no fault is named EXP34-3-nf. You may want to open this file to compare your results with the computer simulation. Then open each of the files with faults. Use the simulated instruments to investigate the circuit and determine the problem. The following are the filenames for circuits with troubleshooting problems for this experiment.

EXP34-3-f1
    Fault: _____

EXP34-3-f2
    Fault: _____

EXP34-3-f3
    Fault: _____

35 Field-Effect Transistors

Name _____
Date _____
Class _____

READING
Text, Section 17–5

OBJECTIVES
After performing this experiment, you will be able to:
1. Measure and graph the drain characteristic curves for a junction field-effect transistor (JFET).
2. Use the characteristic drain curves to determine the transconductance of the JFET.
3. Explain how a JFET can be used as a two-terminal constant-current source.

MATERIALS NEEDED
Resistors:

One 100 Ω, one 33 kΩ

One 2N5458 N-channel JFET transistor (or equivalent)

SUMMARY OF THEORY
The field-effect transistor (FET) is a voltage-controlled transistor that uses an electrostatic field to control current rather than a base current. Instead of a sandwich of materials as in the bipolar junction transistor, the FET begins with a doped piece of silicon called a *channel*. On one end of the channel is a terminal called the *source* and on the other end of the channel is a terminal called the *drain*. Current in the channel is controlled by a voltage applied to a third terminal called the *gate*. Field-effect transistors are classified as either junction-gate (JFET) or insulated-gate (IGFET) devices. Insulated gate devices are also called MOSFETs (for *M*etal *O*xide *S*emiconductor FETs). The major difference between bipolar and field-effect transistors is that BJTs use a small base *current* to control a larger current, but the FET uses a gate *voltage* to control the current. Since the input of a FET draws virtually no current, the input impedance is extremely high; however, the sensitivity to input voltage change is much greater in the bipolar junction transistor than in the FET. Both the JFET and MOSFET have similar ac characteristics; however, in this experiment, we will concentrate on the JFET to simplify the discussion.

The gate of a JFET is made of the opposite type of material from that of the channel, forming a PN diode between the gate and channel. Application of a reverse bias on this junction decreases the conductivity of the channel, reducing the source-drain current. The gate diode is never forward-biased and hence draws almost no current. The JFET comes in two forms, N-channel and P-channel. The N-channel is distinguished on drawings by an inward drawn arrow on the gate connection, while the P-channel has an outward pointing arrow on the gate.

The characteristic drain curves for a JFET exhibit several important differences from the BJT. Besides being a voltage-controlled device, the JFET is a normally ON device. In other words, a reverse-bias voltage must be applied to the gate-source diode in order to close off the channel and stop drain-source current. When the gate is shorted to the source, there is a maximum allowable drain-source current. This current is called I_{DSS} for *D*rain-*S*ource current with gate *S*horted. The JFET exhibits a region on its characteristic curve where drain current is proportional to the drain-source voltage. This region, called the ohmic region, has important applications as a voltage-controlled resistance.

A useful specification for estimating the gain of a JFET is called the *transconductance*, which is abbreviated g_m. Recall that conductance is the reciprocal of resistance. Since the output current is controlled by an input voltage, it is useful to think of FETs as transconductance amplifiers. The transconductance can be found by dividing a small change in the *output current* by a small change in the *input voltage;* that is,

$$g_m = \frac{\Delta I_D}{\Delta V_{GS}}$$

PROCEDURE

1. Measure and record the resistance of the resistors listed in Table 35–1.

2. Construct the circuit shown in Figure 35–1. Start with V_{GG} and V_{DD} at zero volts. Connect a voltmeter between the drain and source of the transistor. Keep V_{CC} at 0 V and slowly increase V_{DD} until V_{DS} is 1.0 V. (V_{DS} is the voltage between the transistor's drain and source.)

Table 35–1

	Listed Value	Measured Value
R_1	33 kΩ	
R_2	100 Ω	

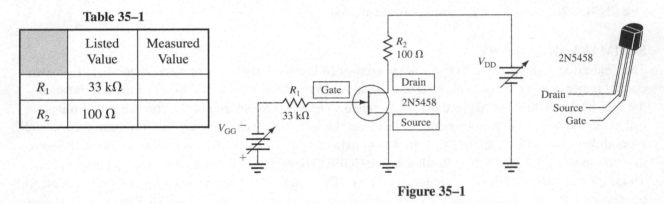

Figure 35–1

3. With V_{DS} at 1.0 V, measure the voltage across R_2 (V_{R2}). Compute the drain current, I_D, by applying Ohm's law to R_2. Note that the current in R_2 is the same as I_D for the transistor. Enter the computed I_D in Table 35–2 under the columns labeled Gate Voltage = 0 V.

4. Without disturbing the setting of V_{GG}, slowly increase V_{DD} until V_{DS} is 2.0 V. Then measure and record V_{R2} for this setting. Compute I_D as before and enter the computed current in Table 35–2 under the columns labeled Gate Voltage = 0 V.

Table 35–2

V_{DS} (measured)	Gate Voltage = 0 V		Gate Voltage = −1.0 V		Gate Voltage = −2.0 V	
	V_{R2} (measured)	I_D (computed)	V_{R2} (measured)	I_D (computed)	V_{R2} (measured)	I_D (computed)
1.0 V						
2.0 V						
3.0 V						
4.0 V						
6.0 V						
8.0 V						

5. Repeat step 4 for each value of V_{DS} listed in Table 35–2.

6. Adjust V_{GG} for −1.0 V. This applies −1.0 V between the gate and source because there is almost no gate current into the JFET and almost no voltage drop across R_1. Reset V_{DD} until V_{DS} = 1.0 V. Measure V_{R2} and enter it in Table 35–2. Compute I_D and enter the computed current in Table 35–2 under the columns labeled Gate Voltage = −1.0 V.

7. Without changing the setting of V_{GG}, adjust V_{DD} for each value of V_{GS} listed in Table 35–2 as before. Compute the drain current at each setting and enter it in Table 35–2 under the columns labeled Gate Voltage = −1.0 V.

8. Adjust V_{GG} for −2.0 V.[1] Repeat steps 6 and 7, entering the data under the columns labeled Gate Voltage = −2.0 V.

9. The data in Table 35–2 represent three drain characteristic curves for your JFET. The drain characteristic curve is a graph of V_{DS} versus I_D for a constant gate voltage. Plot the three drain characteristic curves on Plot 35–1. Choose a scale for I_D that allows the largest current observed to fit on the graph. Label each curve with the gate voltage it represents.

---

[1]The gate-source cutoff voltage for the 2N5458 can vary from −1.0 V to −7.0 V. You may find that −2.0 V turns off the transistor. If the transistor is turned off, try testing it with a gate voltage of −0.5 V.

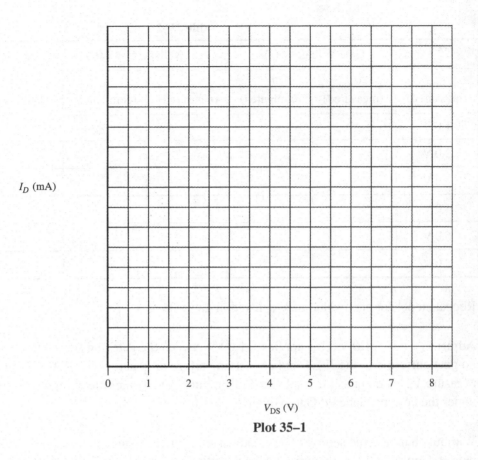

I_D (mA)

V_{DS} (V)

Plot 35–1

10.   Determine the approximate transconductance (g_m) of your JFET at $V_{DS} = 6$ V. Do this by observing the change in drain current between two of the characteristic curves at $V_{DS} = 6$ V and dividing it by a change in the gate-source voltage. Note that the change in the gate-source voltage is 1.0 V between each plotted curve. You should be able to find a transconductance that agrees with the specified range for the JFET you are using, typically 1000 μS to several thousand μS.

$g_m =$

CONCLUSION

EVALUATION AND REVIEW QUESTIONS
1.   (a)   Explain how to find I_{DSS} from the characteristic curves of a JFET.

      (b)   From your data, what is the I_{DSS} for your JFET?

2.   Using the data when the gate voltage is 0 V, explain how you could use your JFET as a two-terminal current source that gives a current of I_{DSS}.

3.   (a)   Does the experimental data indicate that the transconductance is a constant at all points?

      (b)   From your experimental data, what evidence indicates that a JFET is a nonlinear device?

4.   Look up the meaning of pinch-off voltage when $V_{GS} = 0V$. Note that the *magnitude* of V_{GS} is equal to the *magnitude* of V_p, so we can use the characteristic curve for $V_{GS} = 0$ to determine V_p. Using the data from this experiment, determine the pinch-off voltage for your JFET.

5.   Why should a JFET be operated with only reverse bias on the gate source?

FOR FURTHER INVESTIGATION
Using the test circuit shown in Figure 35–1, test the effect of varying V_{GS} with V_{DD} held at a constant +10 V. Tabulate a set of data of I_D as a function of V_{GS} (Table 35–3). Start with $V_{GS} = 0.0$ V and take data every −5.0 V until there is no appreciable drain current. Then graph the data on Plot 35–2. This curve is the transconductance curve for your JFET. The data you obtain are nonlinear because the gate-source voltage is proportional to the square root of the drain current. To illustrate this, compute the square root of I_D and plot the square root of the drain current as a function of the gate-source voltage on Plot 35–3.

Table 35–3

V_{GS} (measured)	I_D (measured)	(computed)
0.0 V		
−0.5 V		
−1.0 V		
−1.5 V		
−2.0 V		
−2.5 V		
−3.0 V		
−3.5 V		
−4.0 V		
−4.5 V		
−5.0 V		

I_D (mA)

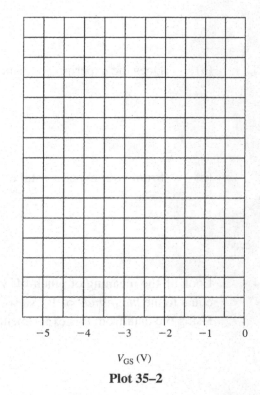

V_{GS} (V)

Plot 35–2

$\sqrt{I_D}$ ($\sqrt{mA}$)

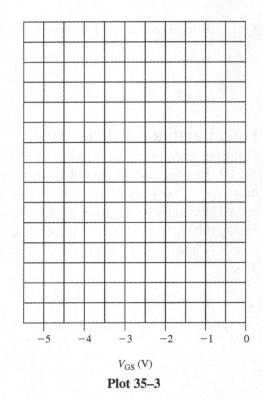

V_{GS} (V)

Plot 35–3

36 Feedback Oscillators

Name _____
Date _____
Class _____

READING
Text, Section 17–7

OBJECTIVES
After performing this experiment, you will be able to:
1.   Connect a class A amplifier; calculate and measure the dc and ac parameters.
2.   Modify the amplifier with a feedback circuit that forms two versions of *LC* oscillators—the Colpitts and the Hartley.
3.   Compare the computed and measured performance of the oscillators.

MATERIALS NEEDED
2N3904 NPN transistor (or equivalent)
One 100 Ω potentiometer
Resistors:
      One 1.0 kΩ, one 2.7 kΩ, one 3.3 kΩ, one 10 kΩ
Capacitors:
      Two 1000 pF, one 0.01 μF, three 0.1 μF
Inductors:
      One 2 μH (can be wound quickly from #22 wire), one 25 μH
For Further Investigation:
      One 1 MHz crystal, one 2N5458 N-channel JFET, one 10 MΩ resistor

SUMMARY OF THEORY
In electronic systems, there is almost always a requirement for one or more circuits that generate a continuous waveform. The output voltage can be a square wave, sine wave, sawtooth, or other periodic waveform. A free-running oscillator is basically an amplifier that generates a continuous alternating voltage by feeding a portion of the output signal back to the input. This type of oscillator is called a feedback oscillator; two types will be investigated in this experiment. A second type of oscillator is called a relaxation oscillator. Relaxation oscillators typically charge a capacitor to some point, then switch it to a discharge path. They are useful for generating sawtooth, triangle, and square waves. A relaxation oscillator (using a 555 timer IC) will be investigated in Experiment 43.

    Feedback oscillators are classified by the networks used to provide feedback. To sustain oscillations, the amplifier must have sufficient gain to overcome the losses in the feedback network. In addition, the feedback must be of the proper phase to ensure that the signal is reinforced at the output—in other words, there must be *positive* feedback. Feedback networks can be classified as *LC, RC,* or by a *crystal,* a special piezoelectric resonant network.

    LC circuits have a parallel resonant circuit, commonly referred to as the *tank* circuit, that determines the frequency of oscillation. A portion of the output is returned to the input causing the amplifier to conduct only during a very small part of the total period. This means that the amplifier is actually run in class C mode. *LC* circuits are generally preferred for frequencies above 1 MHz, whereas *RC* oscillators are usually limited to frequencies below 10 MHz, where stability is not as critical. In applications where frequency stability is important, crystal oscillators have the advantage. In this experiment, you will test two *LC* oscillators, and in the For Further Investigation section, you will test a crystal oscillator. Later, in Experiment 41, you will test a low-frequency *RC* oscillator, the Wien bridge oscillator.

PROCEDURE

1. Measure and record the value of the resistors listed in Table 36–1.

2. Observe the class A amplifier shown in Figure 36–1. Using your measured resistor values, compute the dc parameters for the amplifier listed in Table 36–2. R_{E1} is a 100 Ω potentiometer that you should set to 50 Ω. Then construct the circuit and verify that your computed dc parameters are as expected.

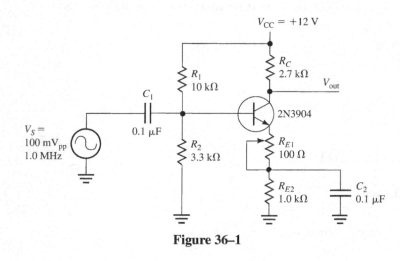

Figure 36–1

Table 36–1

Resistor	Listed Value	Measured Value
R_1	10 kΩ	
R_2	3.3 kΩ	
R_{E1}	50 Ω*	
R_{E2}	1.0 kΩ	
R_C	2.7 kΩ	

*Set potentiometer for 50 Ω

3. Compute the ac parameters listed in Table 36–3. After you find r_e, find the gain by dividing the collector resistance by the sum of the unbypassed emitter resistance and r_e. (Assume that the potentiometer remains set to 50 Ω.) Find the ac voltage at the collector by multiplying the gain by the ac base voltage. Set the generator for a 100 mV$_{pp}$ sine wave at 1.0 MHz and measure the peak-to-peak collector voltage. Your computed quantities should agree with the measured values within experimental uncertainty. Record the measured ac parameters in Table 36–3.

Table 36–2

DC Parameter	Computed Value	Measured Value
V_B		
V_E		
I_E		
V_C		

Table 36–3

AC Parameter	Computed Value	Measured Value
V_b	100 mV$_{pp}$	
r_e		
A_v		
V_c		

4. Remove the generator and add the feedback network for a Colpitts oscillator, as shown in Figure 36–2. Adjust R_{E1} for the best output sine wave. Compute the frequency of the Colpitts oscillator and record the computed frequency in Table 36–4. Then, measure the frequency and the peak-to-peak voltage V_{pp} at the output and record them in Table 36–4.

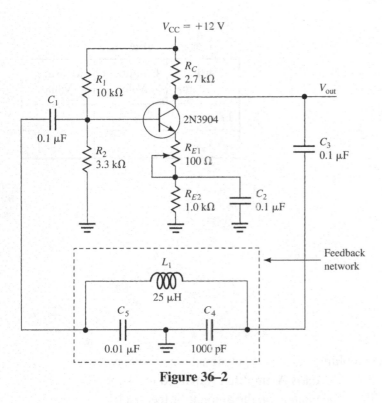

Figure 36–2

Table 36–4

Colpitts Oscillator	Computed Value	Measured Value
Frequency		
V_{pp}		

5. Observe what happens to the frequency and amplitude of the output signal when another 1000 pF capacitor is placed in parallel with C_4.

6. Observe the effect of freeze spray on the stability of the oscillator.

7. Replace the feedback network with the one shown in Figure 36–3. (L_2 can be wound by wrapping about 40 turns of #22 wire on a pencil.) Adjust R_{E1} for a good sine wave output. This configuration is that of a Hartley oscillator. Compute the frequency of the Hartley oscillator, and record the computed frequency in Table 36–5. Then, measure the frequency and the peak-to-peak voltage at the output and record them in Table 36–5.

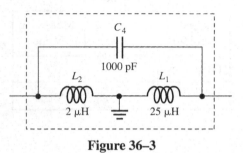

Figure 36–3

Table 36–5

Hartley Oscillator	Computed Value	Measured Value
Frequency		
V_{pp}		

CONCLUSION

EVALUATION AND REVIEW QUESTIONS

1. In step 5, you observed a change in the amplitude of the output signal when a capacitor was placed in parallel with C_4. Since the gain of the class A amplifier remained the same, what conclusion can you draw about the effect of the change on the amount of feedback?

2. What are the two conditions required for oscillation to occur in an *LC* oscillator?

3. Give a reason that an oscillator might drift from its normal frequency.

4. Summarize the difference between a Colpitts and a Hartley oscillator.

5.   For the circuit in Figure 36–2, predict the outcome in each case.

(a) R_{E1} is shorted.

(b) C_4 and C_5 are reversed.

(c) C_2 is open.

(d)   The power supply voltage is 6 V.

FOR FURTHER INVESTIGATION

When it is necessary to have high stability in an oscillator, the crystal oscillator is superior. For high-frequency crystal oscillators, FETs have advantages over bipolar transistors because of their high input impedance. This allows the tank circuit to be unloaded, resulting in a high Q. The circuit shown in Figure 36–4 has the advantage of being simple, yet very stable. Construct the circuit and observe the waveform at the drain. Compare the frequency with that stamped on the crystal case (to do this requires a frequency counter). Test the effect of freeze spray on the frequency and amplitude. Summarize your results.

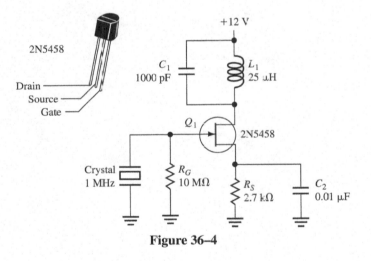

Figure 36–4

299

Application Assignment 17

Name _____

Date _____

Class _____

REFERENCE

Text, Chapter 17; Application Assignment: Putting Your Knowledge to Work

Step 1   Develop a schematic of the circuit from the PC board layout and identify the type of biasing. Draw the schematic in the space provided.

Step 2   Analyze the circuit in the controlled range. With $V_{CC} = +15$ V, determine the output for each temperature listed in Table AA–17–1.

Table AA–17–1

Temperature	Thermistor Resistance	V_{out} (computed)
50° C	2.75 kΩ	
49° C	3.10 kΩ	
51° C	2.50 kΩ	

Step 3   Analyze the circuit in the range from 30° C to 110° C. With $V_{CC} = +15$ V, determine the output for each temperature listed in Table AA–17–2. You will need to determine the thermistor resistance from the graph.

Table AA–17–2

Temperature	Thermistor Resistance	V_{out} (computed)
30° C		
50° C	2.75 kΩ	
70° C		
90° C		
110° C		

Step 4   Troubleshoot the circuit. Give the likely problem(s) for each symptom. Describe how you will isolate the problem.

1. V_{CE} is approximately 0.1 V and V_C is 3.8 V.

_____

_____

2. Collector of Q_1 remains at approximately +15 V.

_____

_____

RELATED EXPERIMENT

MATERIALS NEEDED

One small-signal NPN transistor: 2N3904 (or equivalent)
Resistors:
       One 330 Ω, one 1.0 kΩ, one 22.0 kΩ
One 10 kΩ potentiometer

DISCUSSION

Assume you needed to test the circuit given in the text application assignment. Temperature changes can be simulated by replacing the thermistor with a 10 kΩ potentiometer. Connect the circuit on a breadboard and set the potentiometer to each resistance listed in Tables AA–17–1 and AA–17–2. Add a column to each table for the measured output voltage and record your data. Write a conclusion summarizing your observations.

Checkup 17

Name _____
Date _____
Class _____

REFERENCE

Text, Chapter 17; Lab manual, Experiments 33 through 36

1. To bias a PNP transistor, the polarity of the base voltage with respect to the emitter is:
    (a) negative on both the base and collector
    (b) negative on the base, positive on the collector
    (c) positive on both the base and collector
    (d) positive on the base, negative on the collector

2. When a bipolar transistor is in saturation, the voltage between the collector and emitter is closest to:
    (a) 0.1 V   (b) 0.7 V   (c) 15 V   (d) equal to the supply voltage

3. In order to saturate a bipolar transistor, the base current must be:
    (a) 0   (b) 0.1 mA   (c) greater than I_C (d) greater than I_C divided by β

4. The transconductance of a JFET is found by dividing a change in drain current by a:
    (a) change in source voltage   (b) change in gate voltage
    (c) change in drain voltage   (d) change in gate current

5. Assume an N-channel JFET is operated above pinch-off with V_{GS} equal to zero volts. The drain current is:
    (a) zero   (b) equal to the supply voltage divided by R_D
    (c) I_{DSS} (d) not able to be determined

6. An effective way to increase the gain in a common-emitter amplifier is:
    (a) bypass the emitter resistor with a capacitor
    (b) increase the bias resistors
    (c) bypass the collector resistor with a capacitor
    (d) increase V_{CC}

7. Assume an emitter bypass capacitor is open in a common-emitter amplifier. This would cause:
    (a) a change in the gain
    (b) a change in r_e
    (c) decrease in the input impedance
    (d) all of these

8.   Compared to a common-emitter amplifier, an advantage of a common-source FET amplifier is greater:
     (a) voltage gain   (b) current gain   (c) linearity   (d) input impedance

9.   An *LC* oscillator circuit must have:
     (a) a crystal
     (b) current gain
     (c) a bipolar transistor
     (d) positive feedback

10. The frequency of an *LC* oscillator can be changed by varying:
     (a) the bias
     (b) the emitter bypass capacitor
     (c) components in the feedback path
     (d) the emitter resistance

11. Assume the ac collector voltage of a common-emitter amplifier shows a sinusoidal waveform that is clipped on top. Is this saturation clipping or cutoff clipping? Explain your answer.

12. For the circuit shown in Figure C–17–1 assume that $V_S = +1.4$ V.
     (a) What is the drain voltage? _____
     (b) What is the drain-source voltage? _____
     (c) What type of bias is this? _____

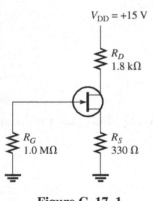

$V_{DD} = +15$ V

R_D
1.8 kΩ

R_G
1.0 MΩ

R_S
330 Ω

Figure C–17–1

13. For the circuits shown in Figure C–17–2 and C–17–3, compare the input resistance of the bipolar junction transistor amplifier with the JFET amplifier.

14. Calculate the gain of the CE amplifier in Figure C–17–2.

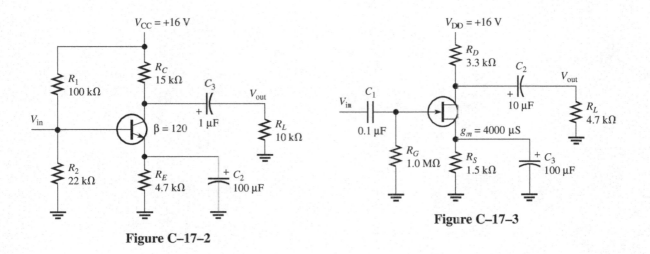

Figure C–17–2

Figure C–17–3

15. For the JFET amplifier in Figure C–17–3, assume that g_m is 4000 μS. Compute the gain.

16. Assume a feedback network in a Hartley oscillator returns 5% of the signal to the input. What is the minimum gain of the amplifier to sustain oscillation?

37 The Differential Amplifier

Name _____

Date _____

Class _____

READING

Text, Sections 18–1 and 18–2

OBJECTIVES

After performing this experiment, you will be able to:
1.   Compute dc and ac parameters of a differential amplifier with emitter bias.
2.   Build a differential amplifier circuit, test it, and measure differential and common-mode gain.

MATERIALS NEEDED

Resistors:

    Two 47 Ω, two 10 kΩ, two 47 kΩ

Two 1 μF capacitors

Two 2N3904 NPN transistors (or equivalent)

For Further Investigation:

    Two 10 kΩ resistors, one 4.7 kΩ resistor, one 2N3904 transistor

SUMMARY OF THEORY

The differential amplifier (*diff-amp*) is used to amplify the *difference* in two signals. The output is related only to the difference between the two inputs. If the input signals are identical, the operation is said to be *common-mode*. The output of the ideal difference amplifier will be zero for common-mode signals. When the inputs are different, the operation is said to be *normal-mode*. The normal-mode signal is amplified by the differential amplifier. The difference amplifier is important in applications in which a weak signal is in the presence of unwanted noise as illustrated in Figure 37–1. The desired transducer signal drives the diff-amp in normal mode. The noise source will tend to induce equal voltages into the twisted pair wires, driving the diff-amp in common-mode. The signal at the output will represent the original transducer signal without the induced noise voltage.

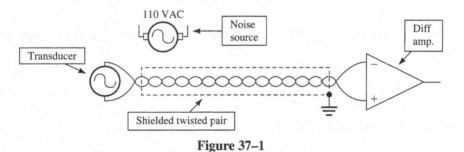

Figure 37–1

The diff-amp's ability to reject unwanted common-mode signals while favoring the desired normal mode signals is called the amplifier's *common-mode rejection ratio* (CMRR). CMRR is often expressed in decibels (dB) and can be defined as:

$$CMMR = 210 \log\left(\frac{A_{v(d)}}{A_{cm}}\right)$$

In this equation, $A_{v(d)}$ represents the differential voltage gain and A_{cm} represents the common-mode voltage gain. The CMRR is a dimensionless number.

PROCEDURE

1.   Measure and record the resistance of the resistors listed in Table 37–1.

Table 37–1

Resistor	Listed Value	Measured Value
R_{B1}	47 kΩ	
R_{B2}	47 kΩ	
R_{E1}	47 Ω	
R_{E2}	47 Ω	
R_T	10 kΩ	
R_C	10 kΩ	

Table 37–2

DC Parameter	Computed Value	Measured Values	
		Q_1	Q_2
I_E			
I_B			
V_B			
V_E			
V_A			
$V_{C(Q2)}$			

2.   The circuit shown in Figure 37–2 is a differential amplifier with single-ended output and emitter bias. Except for V_C, the dc parameters should be identical for Q_1 and Q_2. The transistors are forward biased by the negative supply connected to the common *tail* resistor, R_T. Notice that R_T has *two* emitter currents in it, one from each transistor. To solve for the dc parameters in the circuit, the first step is to write Kirchhoff's voltage equation for the closed path indicated by the dotted line on Figure 37–2. The first sign of each voltage drop is used in writing the equation. (Note that Kirchhoff's voltage law can be written in either direction and obtain the same result.)

$$+V_{EE} - 2I_E R_T - I_E R_{E1} - 0.7 \text{ V} - I_B R_{B1} = 0$$

The base current can be expressed in terms of the emitter current by using the approximation:

$$I_B \cong \frac{I_E}{\beta_{dc}}$$

By substitution, and solving for I_E:

$$I_E = \frac{V_{EE} - 0.7 \text{ V}}{\dfrac{R_{B1}}{\beta_{dc}} + R_{E1} + 2R_T}$$

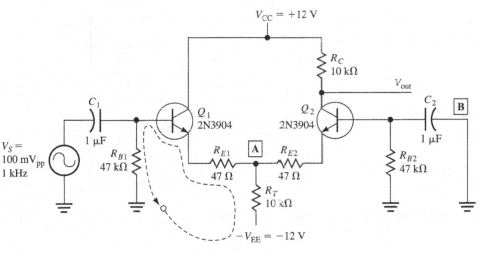

Figure 37–2

Compute I_E for the circuit shown in Figure 37–2. If you do not know the β_{dc} for your transistor, assume a value of 100. Enter your computed I_E in Table 37–2.

3. Compute the remaining dc parameters listed in Table 37–2. The base voltage for either transistor, V_B, can be found by subtracting $I_E R_{B1}$ from zero (ground). The emitter voltage for either transistor is 0.7 V less than V_B. The voltage at point **A** can be computed by subtracting $I_E R_{E1}$ from V_E. $V_{C(Q2)}$ is found by subtracting $I_E R_C$ from V_{CC}.

4. Construct the diff-amp shown in Figure 37–2. The signal generator should be turned off. Measure and record the dc voltages listed in Table 37–2. Your measured and computed values should agree within 10%.

5. Using the computed I_E, calculate r_e for the transistors. Enter the computed r_e in Table 37–3.

6. Compute the differential gain, $A_{v(d)}$, for the circuit. The diff-amp can be thought of as a common-collector amplifier (Q_1) driving a common-base amplifier (Q_2). Point **A** represents the output of the CC amplifier and the input of the CB amplifier. The Q_1 base voltage, $V_{b(Q1)}$, is shown as equal to V_S. The gain to point **A** is approximately ½, so $V_{A(ac)}$ is shown as a computed 50 mV$_{pp}$. The differential gain is the product of the gain of the CC amplifier and the CB amplifier.

$$A_{v(d)} = A_{CC} A_{CB} = \left(\frac{1}{2}\right)\left(\frac{R_C}{R_{E2} + r_{e(Q2)}}\right) = \frac{R_C}{2(R_{E2} + r_{e(Q2)})}$$

Table 37–3

AC Parameter	Computed Value	Measured Value
r_e		
$V_{b(Q1)}$	100 mV$_{pp}$	
$V_{A(ac)}$	50 mV$_{pp}$	
$V_{v(d)}$		
$V_{c(Q2)}$		
A_{cm}		

7. Compute the common-mode gain from the formula

$$A_{cm} \cong \frac{R_C}{2R_T}$$

Enter the computed A_{cm} in Table 37–3.

8. Turn on the signal generator and set V_S for 100 mV$_{pp}$ at 1.0 kHz. Use the oscilloscope to set the proper voltage and check the frequency. (It may be necessary to attenuate the generator input to obtain a 100 mV signal.) Measure the ac signal voltage at Q_1's base ($V_{b(Q1)}$), point **A,** and at Q_2's collector ($V_{c(Q2)}$). Determine the differential gain, $A_{v(d)}$. (The overall voltage gain is the ratio of the output to the input voltage.) Using two channels, observe the phase relationship between these waveforms. Enter your measured results in Table 37–3.

9. To measure the common-mode gain, it is necessary to put an identical signal into both Q_1 and Q_2. Do this by removing point **B** from ground and connecting point **B** to the signal generator. Increase V_S to 1.0 V$_{pp}$ and measure V_{out}. Enter the measured A_{cm} in Table 37–3.

CONCLUSION

EVALUATION AND REVIEW QUESTIONS

1. Using the measured differential gain and the measured common-mode gain, compute the CMRR for the differential amplifier. Express your answer in decibels.

2.	In step 2, Kirchhoff's law was used to develop the equation for the emitter current. The second term of the equation contains $2I_E$ in it. Explain.

3.	In step 6, it is stated that the voltage gain to point **A** is approximately one-half. Explain.

4.	Predict the dc voltage at point **A** if the base of Q_2 is open.

5.	Name at least three malfunctions that could account for a dc voltage of $+12$ V on Q_2's collector.

FOR FURTHER INVESTIGATION

Because current sources have a high internal resistance, the CMRR can be improved by substituting the current source shown in Figure 37–3 for R_T. Measure the differential and common-mode gains and compute the CMRR for the diff-amp tested in this experiment. Summarize your measurements including the differential- and common-mode gains.

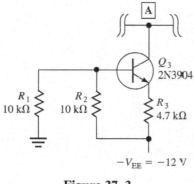

Figure 37–3

MULTISIM TROUBLESHOOTING

This experiment has four Multisim files on the website (www.prenhall.com/floyd). Three of the four files contain a simulated "fault"; one has "no fault". The file with no fault is named EXP37-2-nf. You may want to open this file to compare your results with the computer simulation. Then open each of the files with faults. Use the simulated instruments to investigate the circuit and determine the problem. The following are the filenames for circuits with troubleshooting problems for this experiment.

EXP37-2-f1

    Fault: _____

EXP37-2-f2

    Fault: _____

EXP37-2-f3

    Fault: _____

38 Op-Amp Characteristics

Name _____

Date _____

Class _____

READING
Text, Section 18–3

OBJECTIVES
After performing this experiment, you will be able to:
1. Explain the meaning of common op-amp specifications.
2. Use IC op-amp specification sheets to determine op-amp characteristics.
3. Measure the input offset voltage, bias current, input offset current. and CMRR for a 741C op-amp.

MATERIALS NEEDED
Resistors:

Two 100 Ω, two 10 kΩ, two 100 kΩ, one 1.0 MΩ

Two 1.0 μF capacitors

One 741C op-amp

SUMMARY OF THEORY
An operational amplifier (*op-amp*) is a linear integrated circuit that incorporates a dc-coupled, high-gain differential amplifier and other circuitry that give it specific characteristics. The ideal op-amp has certain unattainable specifications, but hundreds of types of operational amplifiers are available, which vary in specific ways from the ideal op-amp. Important specifications include open-loop gain, input resistance, output resistance, input offset voltage and current, bias current, and slew rate. Other characteristics that are important in certain applications include CMRR, current and voltage noise level, maximum output current, roll-off characteristics, and voltage and power requirements. The data sheet for a specific op-amp contains these specifications, a description of the op-amp, the device pin-out, internal schematic, maximum ratings, suggested applications, and performance curves.

Because the input stage of all op-amps is a differential amplifier, there are two inputs marked with the symbols (+) and (−). These symbols refer to the phase of the output signal compared to the input signal and should be read as noninverting (+) and inverting (−) rather than "plus" or "minus." If the noninverting input is more positive than the inverting input, the output will be positive. If the inverting input is more positive, then the output will be negative. The symbol for an op-amp is shown in Figure 38–1(a). Figure 38–1(b) shows a typical 8-pin dual-in-package (DIP) with an identifier for pin 1. The power supplies are frequently not shown.

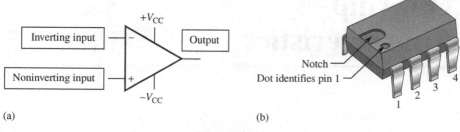

(a)   (b)

Figure 38–1

PROCEDURE

1.   Examine the specification sheet for the 741C op-amp (Appendix C). From the specification sheet, determine the typical and maximum values for each quantity listed in Table 38–1. Record the specified values for $T_A = 25°$ C. Note the measurement units listed on the right side of the specification sheet.

Table 38–1

Step	Parameter	Specified Value			Measured Value
		Minimum	Typical	Maximum	
2d	Input offset voltage, V_{IO}				
3d	Input bias current, I_{BIAS}				
3e	Input offset current, I_{OS}				
4b	Differential gain, $A_{v(d)}$				
4c	Common-mode gain, A_{cm}				
4d	CMRR				

2.   In this step, you will measure the input offset voltage, V_{IO}, of a 741C op-amp. The input offset voltage is the amount of voltage that must be applied between the *input* terminals of an op-amp to give zero *output* voltage.

    (a)   Measure and record the resistors listed in Table 38–2.

    (b)   Connect the circuit shown in Figure 38–2. Install 1 μF bypass capacitor on the power supply leads as shown. Note the polarities of the capacitor.

    (c)   Measure the output voltage, V_{OS}. The input offset voltage is found by dividing the output voltage by the closed-loop gain. The circuit will be considered as a noninverting amplifier for the purpose of the offset calculation.

    (d)   Record the measured input offset voltage in Table 38–1.

Table 38–2

Resistor	Listed Value	Measured Value
R_f	1.0 MΩ	
R_i	10 kΩ	
R_C	10 kΩ	

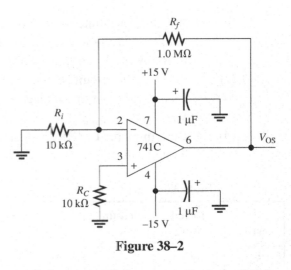

Figure 38–2

3. In this step, you will measure the bias current, I_{BIAS}, and the input offset current, I_{OS}, of a 741C op-amp. The input bias current is the average of the input currents required at each input terminal of the op-amp. The input offset current is a measure of how well these two currents match. The input offset current is the difference in the two bias currents when the output voltage is 0 V.

    (a)   Measure and record the resistors listed in Table 38–3.

    (b)   Connect the circuit shown in Figure 38–3.

    (c)   Measure the voltage across R_1 and R_2 of Figure 38–3. Use Ohm's law to calculate the current in each resistor.

    (d)   Record the *average* of these two currents in Table 38–1 as the input bias current, I_{BIAS}.

    (e)   Record the *difference* in these two currents in Table 38–1 as the input offset current, I_{OS}.

Table 38–3

Resistor	Listed Value	Measured Value
R_1	100 kΩ	
R_2	100 kΩ	

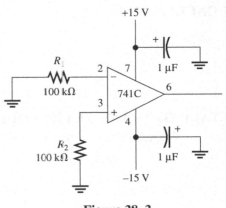

Figure 38–3

4. In this step, you will measure the common-mode rejection ratio, CMRR, of a 741C op-amp. The CMRR is the ratio of the op-amp's differential gain divided by the common-mode gain. It is frequently expressed in decibels according to the definition

$$CMMR = 20 \log \left(\frac{A_{v(e)}}{A_{cm}} \right)$$

    (a)   Measure and record the resistors listed in Table 38–4. For an accurate measurement, resistors R_A and R_B should be closely matched as should R_C and R_D.

315

(b)   It is more accurate to compute the differential gain, $A_{v(d)}$, based on the resistance ratio than to measure it directly. Determine the differential gain by dividing the measured value of R_C by R_A. Enter the differential gain, $A_{v(d)}$, in Table 38–1.

(c)   Connect the circuit shown in Figure 38–4. Set the signal generator for 1.0 V_{pp} at 1 kHz. Measure the output voltage, $V_{out(cm)}$. Determine the common-mode gain, A_{cm}, by dividing $V_{out(cm)}$ by $V_{in(cm)}$.

(d)   Determine the CMRR, in decibels, for your 741C op-amp. Record the result in Table 38–1.

Table 38–4

Resistor	Listed Value	Measured Value
R_A	100 Ω	
R_B	100 Ω	
R_C	100 kΩ	
R_D	100 kΩ	

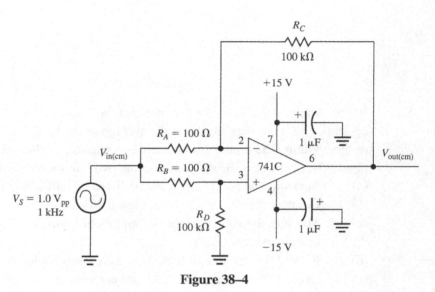

Figure 38–4

CONCLUSION

EVALUATION AND REVIEW QUESTIONS

1.   What is the meaning of the $(+)$ and $(-)$ terminal on the op-amp symbol?

2.   Explain the meaning of input offset voltage.

3.   What is the difference between the input bias current and the input offset current?

4.   What is the difference between differential gain and common-mode gain?

5.   (a)   Explain how you measured the CMRR of the 741C.

        (b)   What is the advantage of a high CMRR?

FOR FURTHER INVESTIGATION

Another important op-amp specification is the *slew rate,* which is defined as the maximum rate of change of the output voltage under large-signal conditions. Slew rate imposes a limit on how fast the output can change and affects the frequency response of the op-amp. It is measured in units of volts/microsecond. It is usually measured using a unity gain amplifier (voltage-follower) with a fast-rising pulse.

        Connect the unity gain circuit shown in Figure 38–5. Set the signal generator for a 10 V_{pp} square wave at 10 kHz. The output voltage will be slew-rate limited. It does not respond instantaneously to the change in the input voltage. Measure the rate of rise (slope) of the output waveform, as shown in Figure 38–6. Compare your measured slew rate with the typical value for a unity-gain 741C op-amp of 0.5 V/μs.

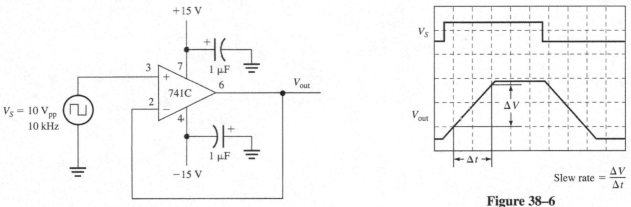

Figure 38–5

$$\text{Slew rate} = \frac{\Delta V}{\Delta t}$$

Figure 38–6

317

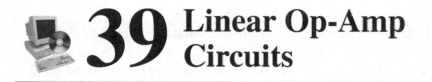

39 Linear Op-Amp Circuits

Name _____
Date _____
Class _____

READING
Text, Sections 18–4 through 18–6

OBJECTIVES
After performing this experiment, you will be able to:
1. Construct and test inverting and noninverting amplifiers using op-amps.
2. Specify components for inverting and noninverting amplifiers using op-amps.

MATERIALS NEEDED
Resistors:
  Two 1.0 kΩ, one 10 kΩ, one 470 kΩ, one 1.0 MΩ
Two 1.0 μF capacitors
One 741C op-amp
For Further Investigation:
  One 1.0 kΩ potentiometer, one 100 kΩ resistor, assorted resistors to test

SUMMARY OF THEORY
One of the most important ideas in electronics incorporates the idea of *feedback,* where a portion of the output is returned to the input. If the return signal tends to decrease the input amplitude, it is called *negative feedback.* Negative feedback produces a number of desirable qualities in an amplifier, increasing its stability and frequency response. It also allows the gain to be controlled independently of the device parameters or changes in temperature and other variables.

Operational amplifiers are almost always used with external feedback. The external feedback network determines the specific characteristics of the amplifier. By itself, an op-amp has an extremely high gain called the *open-loop* gain. When negative feedback is added, the overall gain of the amplifier decreases to an amount determined by the feedback. The overall gain of the amplifier, including the feedback network, is called the *closed-loop* gain, A_{cl}. The closed-loop gain is nearly equal to the reciprocal of the feedback fraction. Figure 39–1 illustrates a noninverting amplifier with negative feedback. The voltage divider samples a fraction of the output voltage and returns it to the inverting input. The closed-loop gain of this noninverting amplifier is given as $A_{cl(NI)}$. It is found by taking the reciprocal of the feedback fraction.

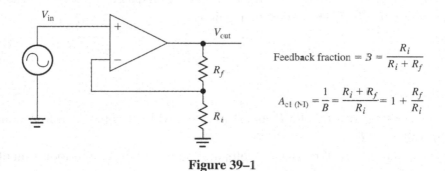

$$\text{Feedback fraction} = \beta = \frac{R_i}{R_i + R_f}$$

$$A_{cl\,(NI)} = \frac{1}{\beta} = \frac{R_i + R_f}{R_i} = 1 + \frac{R_f}{R_i}$$

Figure 39–1

PROCEDURE

1.   The circuit to be tested in this step is the noninverting amplifier illustrated in Figure 39–2. The closed-loop gain equation is given in Figure 39–1.

   (a)   Measure a 10 kΩ resistor for R_f and a 1.0 kΩ resistor for R_i. Record the measured value of resistance in Table 39–1.

   (b)   Using the measured resistance, compute the closed-loop gain of the noninverting amplifier.

   (c)   Calculate V_{out} by multiplying the closed-loop gain by V_{in}.

   (d)   Connect the circuit shown in Figure 39–2. Set the signal generator for a 500 mV$_{pp}$ sine wave at 1.0 kHz. The generator should have no dc offset.

   (e)   Measure the output voltage, V_{out}. Record the measured value.

   (f)   Measure the feedback voltage at pin 2. Record the measured value.

   (g)   Place a 1.0 MΩ resistor in series with the generator. Compute the input resistance of the op-amp based on the voltage drop across the resistor.

Table 39–1

R_f Measured Value	R_i Measured Value	V_{in} Measured	$A_{cl(NI)}$ Computed	V_{out} Computed	V_{out} Measured (pin 6)	$V_{(-)}$ Measured (pin 2)	R_{in} Measured
		500 mV$_{pp}$					

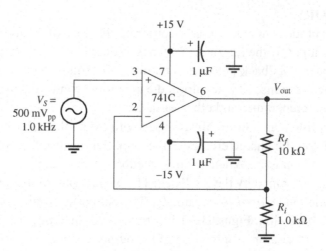

Figure 39–2

2.   In this step you will test an inverting amplifier. Record all data in Table 39–2. The circuit is illustrated in Figure 39–3. The closed-loop gain is

$$A_{cl(I)} = -\left(\frac{R_f}{R_i}\right)$$

   (a)   Use the same resistors for R_f and R_i as in step 1. Record the measured value of resistance in Table 39–2.

   (b)   Using the measured resistance, compute and record the closed-loop gain of the inverting amplifier.

320

(c) Calculate V_{out} by multiplying the closed-loop gain by V_{in}.

(d) Connect the circuit shown in Figure 39–3. Set the signal generator for a 500 mV$_{pp}$ sine wave at 1.0 kHz. The generator should have no dc offset.

(e) Measure and record the output voltage, V_{out}.

(f) Measure and record the voltage at pin 2. This point is called a *virtual ground* because of the feedback.

(g) Place a 1.0 kΩ resistor in series with the generator and R_i. Compute the input resistance of the op-amp based on the voltage drop across the resistor.

Table 39–2

R_f Measured Value	R_i Measured Value	V_{in} Measured	$A_{cl(I)}$ Computed	V_{out} Computed	V_{out} Measured (pin 6)	$V_{(-)}$ Measured (pin 2)	R_{in} Measured
		500 mV$_{pp}$					

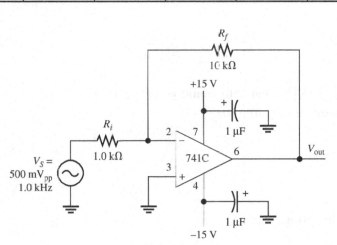

Figure 39–3

3. In this step you will specify the components for an inverting amplifier using a 741C op-amp. The amplifier is required to have an input impedance of 10 kΩ and a closed-loop gain of −47. The input test signal is a 1.0 kHz, 100 mV$_{pp}$ sine wave. (You may need to attenuate your signal generator to obtain this input.) Draw the amplifier. Then build and test your circuit. Find the maximum voltage the input signal can have before clipping occurs. Try increasing the frequency and note the frequency at which the output is distorted. Does the upper frequency response depend on the amplitude of the waveform? Summarize your results in the space provided.

CONCLUSION

EVALUATION AND REVIEW QUESTIONS

1. Express the gain of the amplifiers tested in steps 1 and 2 in decibels.

2. It was not necessary to use any coupling capacitors for the circuits in this experiment. Explain why not.

3. If $R_f = R_i = 10\,\text{k}\Omega$, what gain would you expect for:
   (a) a noninverting amplifier?

   (b) an inverting amplifier?

4. (a) For the noninverting amplifier in Figure 39–2, if $R_f = 0$ and R_i is infinite, what is the gain?

   (b) What is the type of amplifier called?

5.   What output would you expect in the inverting amplifier of Figure 39–3 if R_i were open?

FOR FURTHER INVESTIGATION

An interesting application of an inverting amplifier is to use it as the basis of an ohmmeter for high-value resistors. The circuit is shown in Figure 39–4. The unknown resistor, labeled R_x, is placed between the terminals. The output voltage is proportional to the unknown resistance. Calibrate the meter by placing a known 10 kΩ resistor in place of R_x and adjusting the potentiometer for exactly 100 mV output. The output then represents 10 mV/1000 Ω. By reading the output voltage and moving the decimal point, you can directly read resistors from several thousand ohms to over 1.0 MΩ.

Construct the circuit and test it using different resistors. Calibrate output voltage against resistance and compare with theory. Find the percent error for a 1.0 MΩ resistor using a lab meter as a standard. Summarize your results in a short report.

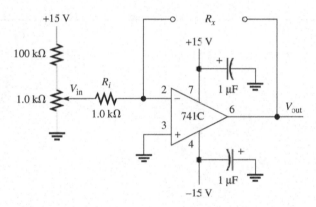

Figure 39–4

MULTISIM TROUBLESHOOTING

This experiment has four Multisim files on the website (www.prenhall.com/floyd). Three of the four files contain a simulated "fault"; one has "no fault". The file with no fault is named EXP39-3-nf. You may want to open this file to compare your results with the computer simulation. Then open each of the files with faults. Use the simulated instruments to investigate the circuit and determine the problem. The following are the filenames for circuits with troubleshooting problems for this experiment.

EXP39-3-f1

    Fault: _____

EXP39-3-f2

    Fault: _____

EXP39-3-f3

    Fault: _____

Application
Assignment 18

Name _____
Date _____
Class _____

REFERENCE

Text, Chapter 18; Application Assignment: Putting Your Knowledge to Work

Step 1   Relate the PC board to a schematic. Draw the schematic in the space provided below:

Step 2   Analyze the circuit.
1.   Determine the resistance value to which the feedback rheostat must be adjusted for a voltage gain of 10. Rheostat setting =
2.   Determine the voltage gain required and the setting of the feedback rheostat for maximum linear output for the specifications given in the text.
$A_v =$ Rheostat setting =
3.   With the gain set as determined in the previous step, on Plot AA–18–1 show the response characteristic of the circuit indicating the output voltage as a function of wavelength between 400 and 700 nm.

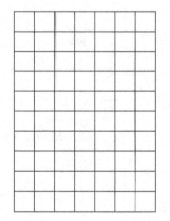

Plot AA–18–1

Step 3   Troubleshoot the circuit. State the probable cause of each of the following problems:
1.   No voltage at the output of the op-amp (list three possible causes).

_____

_____

_____

2.   Output of op-amp stays at approximately -8 V.

_____

3.   A small voltage appears on the output with no light conditions.

_____

RELATED EXPERIMENT

MATERIALS NEEDED
Two 10 kΩ resistors
Two 1 μF capacitors
One 10 kΩ potentiometer
One 741C op-amp

DISCUSSION
As you have seen, the gain of an op-amp can be easily adjusted over a wide range, depending on the requirements. The circuit shown in Figure AA–18–1 is another example of the versatility of op-amps. Can you figure out what the range of gain is? Construct the circuit and test the gain as you vary the potentiometer. Can you classify the amplifier as an inverting type or a noninverting type?

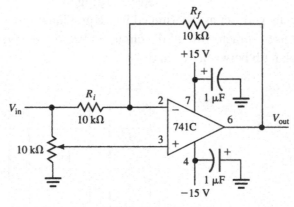

Figure AA–18–1

Checkup 18

Name _____
Date _____
Class _____

REFERENCE

Text, Chapter 18; Lab manual, Experiments 37 through 39

1. A differential amplifier is important in instrumentation systems because it preferentially amplifies:

   (a) both normal- and common-mode signals (b) normal-mode signals only

   (c) common-mode signals only (d) neither

2. Input offset voltage is the voltage applied to the input of an op-amp that will:

   (a) cause the dc output voltage to be zero (b) cause the input current to be zero

   (c) increase the CMRR (d) reduce noise in the output

3. A high CMRR is an advantage for applications requiring:

   (a) high frequency response (b) high gain

   (c) noise rejection (d) input protection

4. The input bias current of an op-amp is:

   (a) the difference in two input currents (b) the average of two input currents

   (c) the current required to produce no output (d) dependent on the gain of the circuit

5. When a portion of output is returned out of phase to the input, it is called:

   (a) regenerative feedback (b) negative feedback

   (c) common-mode feedback (d) closed-loop feedback

6. An ideal op-amp has an open-loop gain of:

   (a) zero (b) 1 (c) 100,000 (d) infinity

7. An ideal op-amp has an input resistance of:

   (a) zero (b) 1 Ω (c) 100,000 Ω (d) infinity

8. An op-amp connected as an inverting amplifier has a 4.7 kΩ input resistor and a 100 kΩ feedback resistor. The input resistance of the amplifier is closest to:

   (a) 4.7 kΩ (b) 22 kΩ (c) 100 kΩ (d) 10 MΩ

9. The amplifier in Question 8 has a voltage gain of approximately:

   (a) 5 (b) 10 (c) 21 (d) 100

10. To measure the input offset current of an op-amp, you could:
    (a) compare the output and input currents in an inverting amplifier
    (b) compare the output and input currents in a noninverting amplifier
    (c) determine the input current both with and without a signal on a diff-amp
    (d) use Ohm's law to determine the current in each of two large input resistors

11. Explain what is meant by a *virtual ground*.

12. Compare the difference between *open-loop* and *closed-loop* gain.

13. Assume you need a noninverting amplifier with a gain of 14.
    (a)   What ratio is required between the feedback resistor and the input resistor?

    (b)   Choose two standard-value resistors that will give the feedback fraction you determined in (a). Draw the circuit and label the resistors.

14. An op-amp can be operated from a single positive power supply, as shown in Figure C–18–1. (The noninverting input has a dc level applied to it to change the reference.) Assume V_{ref} is set for a +6.0 V level. Use the superposition theorem and the ideal op-amp model to answer the following questions.
    (a)   What is the dc level at the inverting input? _____
    (b)   What is the dc voltage across R_i? _____
    (c)   What is the input current? _____
    (d)   What is the current in R_f? _____
    (e)   What is V_{out}? _____

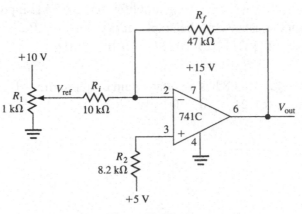

Figure C–18–1

328

40 Nonlinear Op-Amp Circuits

Name _____
Date _____
Class _____

READING
Text, Sections 19–1 through 19–3

OBJECTIVES
After performing this experiment, you will be able to:
1. Construct and test an op-amp comparator, an integrator, and a differentiator circuit.
2. Determine the response of the circuits listed in objective 1 to various waveforms.
3. Troubleshoot faults in op-amp circuits.

MATERIALS NEEDED
Resistors:
> One 330 Ω, one 1.0 kΩ, four 10 kΩ, three 22 kΩ, one 330 kΩ

Capacitors:
> One 2200 pF, one 0.01 μF, two 1.0 μF

Three 741C op-amps
One 1 kΩ potentiometer
Two LEDs (one red, one green)

SUMMARY OF THEORY
The basic op-amp is a linear device; however, many applications exist in which the op-amp is used in a nonlinear circuit. One of the most common nonlinear applications is the comparator. A comparator is used to detect which of two voltages is larger and to drive the output into either positive or negative saturation. Comparators can be made from ordinary op-amps (and frequently are), but there are special ICs designed as comparators. They are designed with very high slew rates and frequently have open-collector outputs to allow interfacing to logic or bus systems.

Other uses of op-amps include a variety of signal processing applications. Op-amps are ideally suited to make precise integrators. Integration is the process of finding the area under a curve, as shown in the Summary of Theory for Experiment 30. An integrator produces an output voltage that is proportional to the *integral* of the input voltage waveform. The opposite of integration is differentiation. Differentiation circuits produce an output that is proportional to the *derivative* of the input voltage waveform. The basic op-amp comparator, integrator, and differentiator circuits with representative waveforms are illustrated in Figure 40–1.

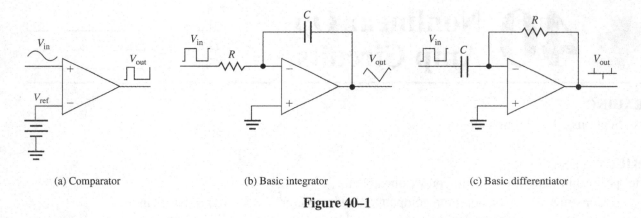

(a) Comparator   (b) Basic integrator   (c) Basic differentiator

Figure 40–1

PROCEDURE

1. In this step you will construct and test an op-amp circuit connected as a comparator. Construct the circuit shown in Figure 40–2. Vary the potentiometer. Measure the output voltage when the red LED is on and then when the green LED is on. The 741C has current-limiting circuitry that prevents excessive current from destroying the LEDs. Record the output voltages, V_{OUT}, in Table 40–1. Notice that the LEDs prevent the output from going into positive and negative saturation. Then set the potentiometer to the threshold point. Measure and record V_{ref} at the threshold.

Table 40–1

V_{OUT}		V_{ref}
Red On	Green On	Threshold

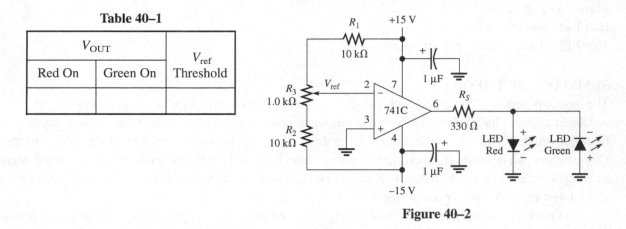

Figure 40–2

2. In this step, you will test the effects of the comparator on a sine wave input and add an integrating circuit to the output of the comparator. Connect the circuit shown in Figure 40–3 and add a sine wave generator to the comparator as illustrated. Set the output for a 1.0 V_{pp} at 1.0 kHz with no dc offset. Observe the waveforms from the comparator (point **A**) and from the integrator (point **B**). Adjust R_3 so that the waveform at **B** is centered about zero volts. Sketch the observed waveforms in the correct time relationship on Plot 40–1. Show the voltages and time on your plot.

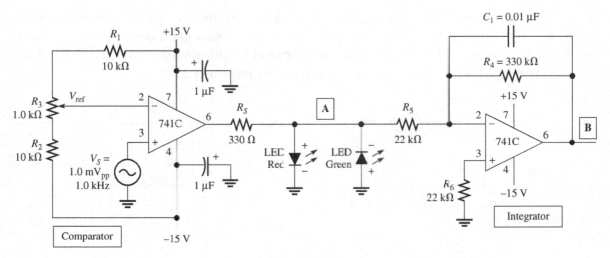

Figure 40–3

Plot 40–1

3. Vary R_3 while observing the output of the comparator and the integrator. Observations:

4. For each of the troubles listed in Table 40–2, see if you can predict the effect on the circuit. Then insert the trouble and check your prediction. At the end of this step, restore the circuit to normal operation.

Table 40–2

Trouble	Symptoms
No negative power supply	
Red LED open	
C_1 open	
R_4 open	

331

5. In this step, you will add a differentiating circuit to the previous circuit. The differentiator circuit is shown in Figure 40–4. Connect the input of the differentiator to the output of the integrator (point **B**). Observe the input and output waveforms of the differentiator. Sketch the observed waveforms on Plot 40–2. Label your plot and show voltage and time.

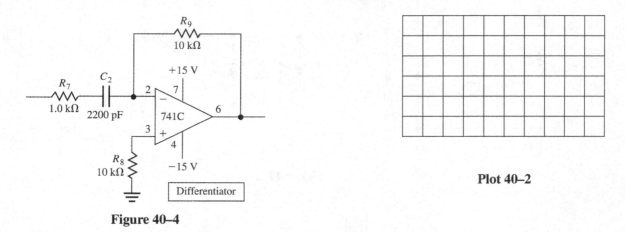

Figure 40–4

Plot 40–2

6. Remove the input from the differentiator and connect it to the output from the comparator (point **A**). Observe the new input and output waveforms of the differentiator. Sketch the observed waveforms on Plot 40–3. Label your plot.

Plot 40–3

CONCLUSION

EVALUATION AND REVIEW QUESTIONS

1. Compute the minimum and maximum V_{ref} for the comparator in Figure 40–2.

    $V_{ref(min)} =$ _____ $V_{ref(max)} =$ _____

2. The comparator output did not go near the power supply voltages. Explain why not.

3. (a) For the integrator circuit in Figure 40–3, what is the purpose of R_4?

(b) What happened when it was removed?

4. What type of circuit will produce leading- and trailing-edge triggers from a square wave input?

5. What effect would you expect on the output of the integrator in Figure 40–3 if the frequency used were 100 Hz instead of 1.0 kHz?

FOR FURTHER INVESTIGATION
A useful variation of the comparator is the Schmitt trigger circuit shown in Figure 40–5. This circuit is basically a comparator that uses *positive* feedback to change the threshold voltage when the output switches. The trip point is dependent on whether the output is already saturated high or low. This effect is called *hysteresis*. Construct the circuit, test its operation, and summarize your findings in a short report.

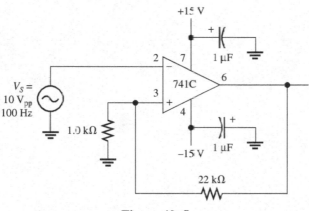

Figure 40–5

MULTISIM TROUBLESHOOTING

This experiment has four Multisim files on the website (www.prenhall.com/floyd). Three of the four files contain a simulated "fault"; one has "no fault". The file with no fault is named EXP40-3-nf. You may want to open this file to compare your results with the computer simulation. Then open each of the files with faults. Use the simulated instruments to investigate the circuit and determine the problem. The following are the filenames for circuits with troubleshooting problems for this experiment.

EXP40-3-f1

       Fault: _____

EXP40-3-f2

       Fault: _____

EXP40-3-f3

       Fault: _____

41 The Wien Bridge Oscillator

Name _____
Date _____
Class _____

READING
Text, Section 19–4

OBJECTIVES
After performing this experiment, you will be able to:
1. Explain the requirements for a Wien bridge to oscillate, predict the feedback voltages and phases, and compute the frequency of oscillation.
2. Construct and test a Wien bridge oscillator with automatic gain control.

MATERIALS NEEDED
Resistors:
     One 1.0 kΩ, three 10 kΩ
Capacitors:
     Two 0.01 μF, three 1 μF
Two 1N914 signal diodes (or equivalent)
One 741C op-amp
One 2N5458 N-channel JFET transistor (or equivalent)
One 10 kΩ potentiometer
For Further Investigation:
   One type 1869 or type 327 bulb

SUMMARY OF THEORY
The Wien bridge is a circuit that is widely used as a sine wave oscillator for frequencies below about 1 MHz. Oscillation occurs when a portion of the output is returned to the input in the proper amplitude and phase to reinforce the input signal. This type of feedback is called *regenerative* or *positive* feedback. Regenerative oscillators require amplification to overcome the loss in the feedback network. For the Wien bridges shown in Figure 41–1, the signal returned to the noninverting input is one-third of the output signal. For this reason, the amplifier must provide a minimum gain of 3 to prevent oscillations from dying out.

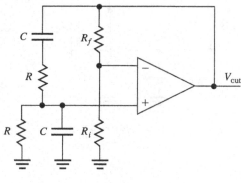

(a) Basic Wien bridge

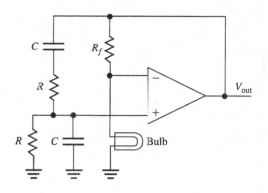

(b) Light bulb stabilized Wien bridge

Figure 41–1

The basic Wien bridge circuit is shown in Figure 41–1(a). The frequency of oscillation is determined by the lead-lag network connected to the noninverting input of the op-amp. The gain is controlled by R_f and R_i. The frequency of oscillation is found from the equation

$$f_r = \frac{1}{2\pi RC}$$

The gain must be at least 3 to maintain oscillations, but too much gain causes the output to saturate. Too little gain causes oscillations to cease. Various circuits have been designed to stabilize loop gain at exactly 3. The basic requirement is to provide *automatic gain control,* or AGC for short. One common technique is to use a light bulb for AGC as illustrated in Figure 41–1(b). As the bulb's filament warms, the resistance increases and reduces the gain. Other more sophisticated techniques use the variable resistance region of a FET to control gain. A FET automatic gain control circuit will be investigated in this experiment.

PROCEDURE

1. Measure R_1, R_2, C_1, and C_2. These components determine the frequency of the Wien bridge. Record the measured values in Table 41–1. If you cannot measure the capacitors, record the listed value.

2. Construct the basic Wien bridge illustrated in Figure 41–2. Adjust R_f so that the circuit just oscillates. You will see that it is nearly impossible to obtain a clean sine wave, as the control is too sensitive. With the bridge oscillating, try spraying some freeze spray on the components and observe the result. Observations:

Table 41–1

Component	Listed Value	Measured Value
R_1	10 kΩ	
R_2	10 kΩ	
C_1	0.01 μF	
C_2	0.01 μF	

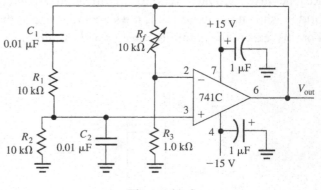

Figure 41–2

3. The basic Wien bridge in step 2 has unstable gain and requires some form of automatic gain control to work properly. Field-effect transistors are frequently used for AGC circuits because they can be used as voltage-controlled resistors for small applied voltages. The circuit illustrated in Figure 41–3 is a FET-stabilized Wien bridge. Compute the expected frequency of oscillation from the equation

$$f_r = \frac{1}{2\pi RC}$$

Use the *average* measured value of the resistance and capacitance listed in Table 41–1 to calculate f_r. Record the computed f_r in Table 41–2.

Table 41–2

f_r	
Computed	Measured (pin 6)

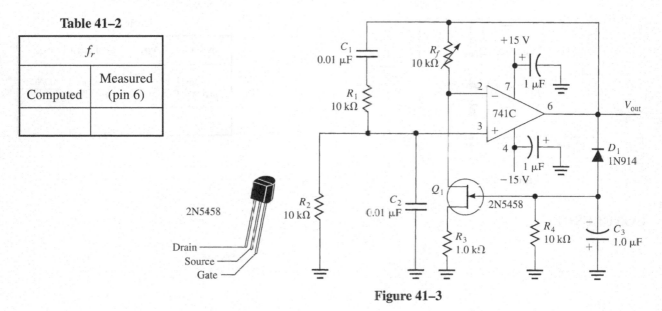

Figure 41–3

4. Construct the FET-stabilized Wien bridge shown in Figure 41–3. The diode causes negative peaks to charge C_3 and bias the FET. C_3 has a long time-constant discharge path, so the bias does not change rapidly. Note the polarity of C_3. Adjust R_f for a good sine wave output. Measure the frequency and record it in Table 41–2.

5. Measure the peak-to-peak output voltage, $V_{out(pp)}$. Then measure the peak-to-peak positive and negative feedback voltages, $V_{(+)(pp)}$ and $V_{(-)(pp)}$ and the dc voltage on the gate of the FET. Use two channels and observe the phase relationship of the waveforms. Record the voltages in Table 41–3.

Table 41–3

Measured Voltages			
$V_{out(pp)}$ (pin 6)	$V_{(+)(pp)}$ (pin 3)	$V_{(-)(pp)}$ (pin 2)	V_{GATE}

What is the phase shift from the output voltage to the positive feedback voltage?

6.   Try freeze spray on the various components while observing the output. Observations:

7.   Add a second diode in series with the first one between the output and the gate of the FET (see Figure 41–4). You may need to readjust R_f for a good sine wave. Measure the voltages as before and record in Table 41–4.

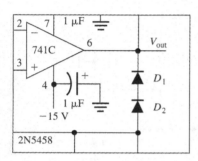

Figure 41–4

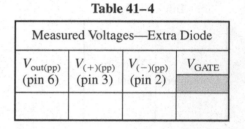

Table 41–4

Measured Voltages—Extra Diode			
$V_{out(pp)}$ (pin 6)	$V_{(+)(pp)}$ (pin 3)	$V_{(-)(pp)}$ (pin 2)	V_{GATE}

CONCLUSION

EVALUATION AND REVIEW QUESTIONS

1.   In step 5, you measured the positive feedback voltage. What fraction of the output voltage did you find? Is this what you expected from theory?

2.   Explain why adding a second diode in series with the first caused the output voltage to increase.

3.   For the circuit in Figure 41–3, why is the positive side of C_3 shown at ground?

4.   At what frequency would the Wien bridge of Figure 41–3 oscillate if R_1 and R_2 were doubled?

5.   How could you make a Wien bridge tune to different frequencies?

FOR FURTHER INVESTIGATION

Investigate the light-bulb-stabilized Wien bridge shown in Figure 41–1(b). A good bulb to try is a type 1869 or type 327. Other bulbs will work, but low-resistance filaments are not good. You can use the same components as in Figure 41–2 except replace R_3 with the bulb. Summarize your results in a short lab report.

MULTISIM TROUBLESHOOTING

This experiment has four Multisim files on the website (www.prenhall.com/floyd). Three of the four files contain a simulated "fault"; one has "no fault". The file with no fault is named EXP41-3-nf. You may want to open this file to compare your results with the computer simulation. Then open each of the files with faults. Use the simulated instruments to investigate the circuit and determine the problem. The following are the filenames for circuits with troubleshooting problems for this experiment.

EXP41-3-f1
    Fault: _____

EXP41-3-f2
    Fault: _____

EXP41-3-f3
    Fault: _____

42 Active Filters

Name _____

Date _____

Class _____

READING
Text, Section 19–5

OBJECTIVES
After performing this experiment, you will be able to:
1. Specify the components required for a Butterworth low- or high-pass filter.
2. Build and test a Butterworth low- or high-pass active filter for a specific frequency.

MATERIALS NEEDED
Resistors:

One 1.5 kΩ, four 8.2 kΩ, one 10 kΩ, one 22 kΩ, one 27 kΩ

Capacitors:

Four 0.01 μF, two 1 μF

Two 741C op-amps

For Further Investigation:

One additional 741C op-amp and components to be specified by student

SUMMARY OF THEORY
A filter is a circuit that produces a prescribed frequency response as described in Experiment 28. Passive filters are combination circuits containing only resistors, inductors, and capacitors (*RLC*). Active filters contain resistance and capacitance plus circuit elements that provide gain, such as transistors or operational amplifiers. The major advantage of active filters is that they can achieve frequency response characteristics that are nearly ideal and for reasonable cost for frequencies up to about 100 kHz. Above this, active filters are limited by bandwidth.

Active filters can be designed to optimize any of several characteristics. These include flatness of the response in the passband, steepness of the transition region, or minimum phase shift. The Butterworth form of filter has the flattest passband characteristic, but is not as steep as other filters and has poor phase characteristics. Since a flat passband is generally the most important characteristic, it will be used in this experiment.

The *order* of a filter, also called the number of *poles*, governs the steepness of the transition outside the frequencies of interest. In general, the higher the order, the steeper the response. The roll-off rate for active filters depends on the type of filter but is approximately −20 dB/decade for each pole. (A *decade* is a factor of 10 in frequency.) A four-pole filter, for example, has a roll-off of approximately −80 dB/decade. A quick way to determine the number of poles is to count the number of capacitors that are used in the frequency-determining part of the filter.

Figure 42–1 illustrates a two-pole active low-pass and a two-pole active high-pass filter. Each of these circuits is a *section*. To make a filter with more poles, simply cascade these sections, but change the gains of each section according to the values listed in Table 42–1. The cutoff frequency will be given by the equation

$$f = \frac{1}{2\pi RC}$$

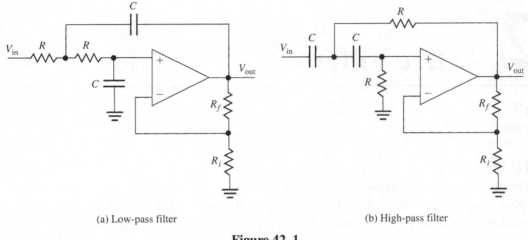

(a) Low-pass filter   (b) High-pass filter

Figure 42–1

You can design your own Butterworth low-pass or high-pass active filter by using the following guidelines:

(a)   Determine the number of poles necessary based on the required roll-off rate. For example, if the required roll-off is −40 dB/decade, specify a two-pole filter.

(b)   Choose R and C values for the desired cutoff frequency. These components are labeled R and C on Figure 42–1. The resistors should be between 1.0 kΩ and 100 kΩ. The values chosen should satisfy the cutoff frequency as given by the equation

$$f = \frac{1}{2\pi RC}$$

(c)   Choose resistors R_f and R_i that give the gains for each section according to the values listed in Table 42–1. The gain is controlled only by R_f and R_i. Solving the closed-loop gain of a noninverting amplifier gives the equation for R_f in terms of R_i:

$$R_f = (A_v - 1)R_i$$

Table 42–1

Poles	Gain Required		
	Section 1	Section 2	Section 3
2	1.586		
4	1.152	2.235	
6	1.068	1.586	2.483

Example:
A low-pass Butterworth filter with a roll-off of approximately −80 dB/decade and a cutoff frequency of 2.0 kHz is required. Specify the components.

Step 1   Determine the number of poles required. Since the design requirement is for approximately -80 dB/decade, a four-pole (two-section) filter is required.

Step 2   Choose R and C. Try C as 0.01 µF and compute R. Computed $R = 7.96$ kΩ. Since the nearest standard value is 8.2 kΩ, choose $C = 0.01$ µF and $R = 8.2$ kΩ.

Step 3   Determine the gain required for each section and specify R_f and R_i. From Table 42–1, the gain of section 1 is required to be 1.152, and the gain of section 2 is required to be 2.235. Choose resistors that will give these gains for a noninverting amplifier. The choices are determined by considering standard values and are shown on the completed schematic, Figure 42–2.

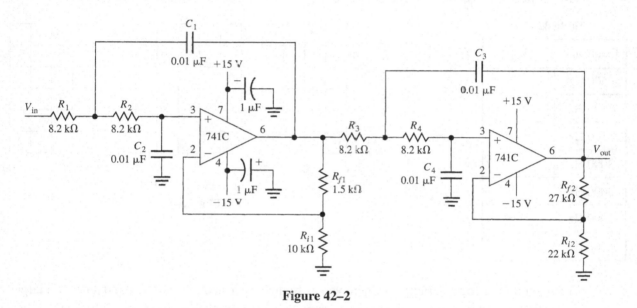

Figure 42–2

PROCEDURE

1.   Measure and record the components listed in Table 42–2. If you cannot measure the capacitors, use the listed value.

Table 42–2

Component	Listed Values	Measured Values			
		1	2	3	4
R_1 to R_4	8.2 kΩ				
C_1 to C_4	0.01 µF				
R_{i1}	10 kΩ				
R_{f1}	1.5 kΩ				
R_{i2}	22 kΩ				
R_{f2}	27 kΩ				

343

2.   Construct the four-pole low-pass active filter shown in Figure 42–2. Install a 10 kΩ load resistor. Connect a generator to the input and set it for a 500 Hz sine wave at 1.0 V_{rms}. The voltage should be measured at the generator with the circuit connected. Set the voltage with a voltmeter and check both voltage and frequency with the oscilloscope. Measure V_{RL} at a frequency of 500 Hz, and record it in Table 42–3.

3.   Change the frequency of the generator to 1000 Hz. Readjust the generator's amplitude to 1.0 V_{rms}. Measure V_{RL}, entering the data in Table 42–3. Continue in this manner for each frequency listed in Table 42–3.

4.   Graph the voltage across the load resistor (V_{RL}) as a function of frequency on Plot 42–1.

Table 42–3

Frequency	V_{RL}
500 Hz	
1000 Hz	
1500 Hz	
2000 Hz	
3000 Hz	
4000 Hz	
8000 Hz	

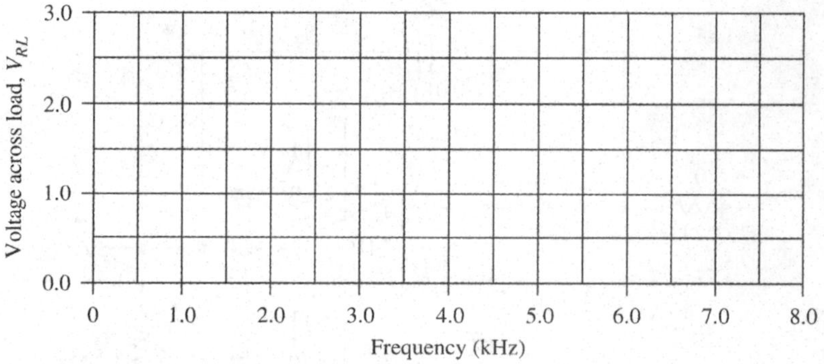

Plot 42–1

5.   A Bode plot is a log-log plot of voltage versus frequency. You can examine the data over a larger range than with linear plots. Plot the data from the filter onto the log-log Plot 42–2. The theoretical roll-off of −80 dB/decade is plotted for reference.

CONCLUSION

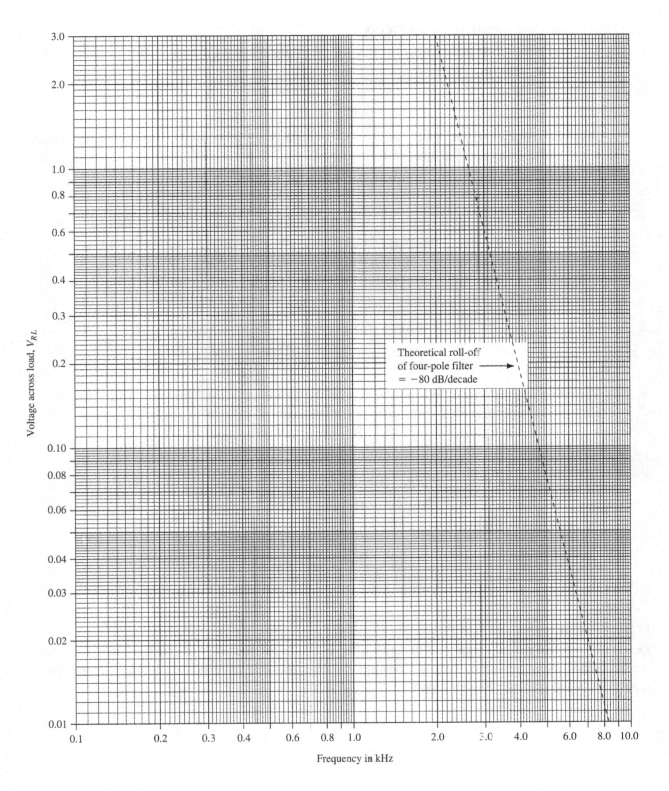

Theoretical roll-off
of four-pole filter ⟶
= −80 dB/decade

Voltage across load, V_{RL}

Frequency in kHz

Plot 42–2

EVALUATION AND REVIEW QUESTIONS

1.   (a)   From the frequency response curves, determine the cutoff frequency for the filter in this experiment.

        (b)   Compute the average R and C for your active filter (Table 42–1). Use the average values of each to compute the cutoff frequency.

2.   The cutoff frequency of the active filter in this experiment is about the same as that of Experiments 20 and 28. Compare the frequency response curves you obtained in those experiments with the response of the active filter.

3.   Using the Bode plot, predict V_{out} at a frequency of 10 kHz.

4.   The reference line on the Bode plot represents the theoretical roll-off rate of -80 dB/decade. How does your actual filter compare to this theoretical roll-off rate?

5.   (a)   Using the measured values of R_{i1} and R_{f1}, compute the actual gain of the first section. Compare this with the required gain in Table 42–1.

        (b)   Repeat for the second section using R_{i2} and R_{f2}.

FOR FURTHER INVESTIGATION

Design a six-pole high-pass Butterworth active filter with a cutoff frequency of 400 Hz. The procedures are outlined in the Summary of Theory. Construct the filter, measure its response, and submit a laboratory report. Show the response in your report.

Name ____________________
Date ____________________
Class ____________________

Application Assignment 19

REFERENCE

Text, Chapter 19; Application Assignment: Putting Your Knowledge to Work

Step 1 Relate the PC board to a schematic. Draw the schematic in the space provided below:

Step 2 Analyze the power supply circuit.

1. Determine the voltage with respect to ground at each of the four "corners" of the bridge. State if the voltage is ac or dc. Complete Table AA–19–1.

Table AA–19–1

Corner:	Voltage and Description
D_1–D_3	
D_2–D_3	
D_1–D_4	
D_2–D_4	

2. Calculate the PIV for the rectifier diodes. PIV =
3. Show the waveform across D_1 for a full cycle of the ac input.

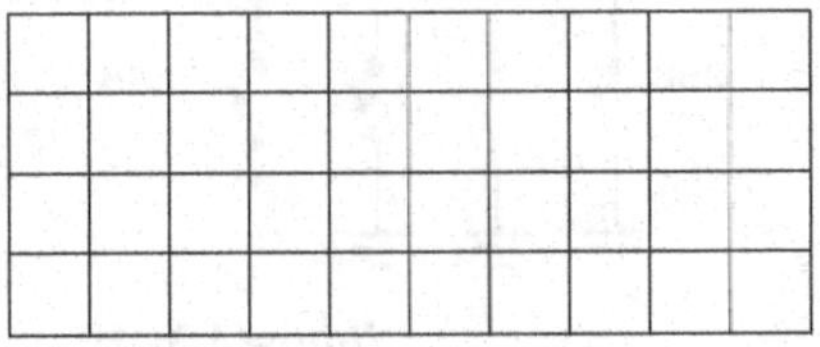

Plot AA–19–1

Step 3   Troubleshoot the power supply for the following problems. State the probable cause.
1.   Both positive and negative outputs are zero: _____
2.   Positive output is zero; negative output is -12 V: _____
3.   Negative output is zero; positive output is $+12$ V: _____
4.   Radical fluctuations on output of positive regulator: _____

Indicate the voltage you would measure for each of the following faults:
1.   Diode D_1 open _____
2.   Capacitor C_2 open _____

RELATED EXPERIMENT

MATERIALS NEEDED
One 741C op-amp
Four 1.0 kΩ resistors; additional resistors as determined by student

DISCUSSION
A scaling adder can be used to convert a binary number into an analog voltage level. Binary numbers employ only the digits 0 and 1 (called bits) to form numbers. Each bit can be represented as either a closed or open switch (0 or 1), as illustrated in Figure AA–19–1. As in any weighted counting system, the digits to the left take on higher "weights" based on their position—in the binary system, the weight of each column is a factor of two larger than the column on the immediate right.

For this related experiment, devise a scaling adder circuit that will convert a 4-bit binary number into analog voltage that is proportional to the input binary number. You may choose any gain you like, but the gain should not be so high as to drive the op-amp into saturation for the largest binary number (1111). The least significant bit will have a weight of 1 and the most significant bit will have a weight of 8. Draw your circuit and include measured values for the sixteen combinations of inputs.

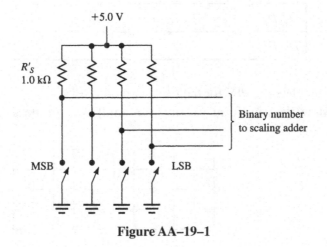

Figure AA–19–1

Checkup 19

REFERENCE
Text, Chapter 19; Lab manual, Experiments 40, 41, and 42

1. To connect an op-amp as a comparator, it is *not* necessary to have:
    (a) power supplies   (b) a reference
    (c) feedback resistors   (d) an input voltage

2. When the input voltage crosses the reference voltage of a comparator, the output:
    (a) changes state   (b) becomes zero   (c) oscillates   (d) goes to +5.0 V

3. When a square wave is the input signal of an integrating circuit, the output is a:
    (a) triangle waveform   (b) sinusoidal waveform
    (c) step waveform   (d) series of positive and negative pulses

4. When a square wave is the input signal of a differentiating circuit, the output is a
    (a) triangle waveform   (b) sinusoidal waveform
    (c) step waveform   (d) series of positive and negative pulses

5. A circuit that can be used to generate a sinusoidal waveform is:
    (a) an integrator   (b) a differentiator
    (c) a Wien bridge   (d) a comparator

6. The gain for a Wien bridge oscillator must be at least:
    (a) 3   (b) 15   (c) 29   (d) depends on the op-amp used

7. The critical frequency of a low-pass filter is the frequency at which the output signal, when compared to the midband level, is attenuated by:
    (a) 0 dB   (b) -3 dB   (c) -6 dB   (d) depends on the number of poles

8. A four-pole Butterworth filter has a roll-off rate of approximately:
    (a) -20 dB per decade   (b) -40 dB per decade
    (c) -60 dB per decade   (d) -80 dB per decade

9. If the resistors and capacitors are interchanged on a low-pass filter, the result is:
    (a) another low-pass filter   (b) a high-pass filter
    (c) a bandpass filter   (d) a notch filter

10. In a series regulator, if the output voltage decreases because the input voltage drops, the voltage on the base of the pass transistor
    (a) increases   (b) decreases   (c) remains the same

11. Assume that a zero-crossing detector has a 10 V_{pp} sine wave on the inverting terminal and ground on the noninverting terminal, as shown in Figure C–19–1. Sketch the input and output waveforms.

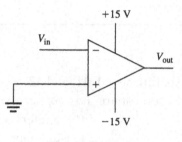

Figure C–19–1

12. Determine the reference voltage for the comparator shown in Figure C–19–2.

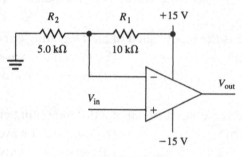

Figure C–19–2

13. Assume two input signals from different microphones are connected to the summing network shown in Figure C–19–3.

(a) What is the gain of the summing amplifier to each input?

(b) Write an expression for the output signal in terms of the two inputs.

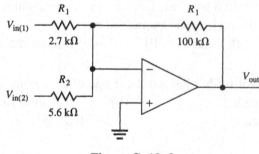

Figure C–19–3

14. (a) Compare the basic operation of an integrating circuit with a differentiating circuit.

(b) Could an integrating or differentiating circuit be constructed using a resistor and inductor? Why do you think that inductors are not used?

43 The Instrumentation Amplifier

Name _____
Date _____
Class _____

READING
Text, Section 20–1

OBJECTIVES
After performing this experiment, you will be able to:
1. Construct an instrumentation amplifier (IA) using three op-amps. Compute the differential gain and measure the common-mode gain. Compute the CMRR' for the IA.
2. Construct an oscillator as an input signal source for the IA constructed in objective 1. Add simulated common-mode noise to the input and adjust the common-mode gain of the amplifier for minimum noise at the output. Demonstrate how an IA can selectively amplify a small differential signal in the presence of a large common-mode signal.

MATERIALS NEEDED
Resistors:
One 22 Ω, one 470 Ω, two 1.0 kΩ, one 8.2 kΩ, five 10 kΩ, three 100 kΩ
Capacitors:
One 0.01 μF, two 1.0 μF
Three LM741C op-amps
One 555 timer
One small 9 V battery
One 5 kΩ potentiometer
30 cm twisted-pair wire
For Further Investigation:
One CdS cell Jameco 120299 (or equivalent)

SUMMARY OF THEORY
Frequently, instrumentation systems have low-level signals from transducers (strain gauges, thermistors, etc.). These signals can be extremely small and may originate from high impedance sources located some distance from the signal conditioning. Noise is frequently a problem with such signals. The requirement for accurate amplifiers for this type of application led to the development of the *instrumentation amplifier,* or IA. Integrated circuit IAs are high-performance, high-impedance amplifiers that use a differential input. Gain is normally set with a single external resistor.

An instrumentation amplifier must also have excellent dc performance characteristics—ideally it should have no dc offset or voltage drift and should reject common-mode noise. Common-mode noise is unwanted voltage that is present on both signal leads and a common reference. Frequently, electronic systems have noise that originates from a common-mode source (typically interference). In order to reject common noise, the differential gain should be high while common-mode gain should be very low (implying a high CMRR). Although most IAs are in IC form, it is possible to construct an IA using conventional op-amps by adding two op-amps to a basic differential amplifier. For this experiment, you will construct an IA using three op-amps in the configuration shown in Figure 43–1. Differential and

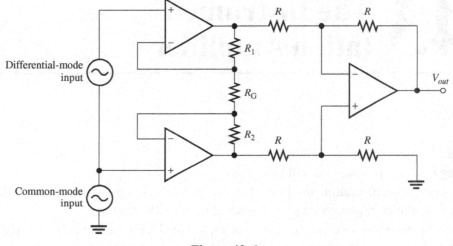

Figure 43–1

common-mode inputs are shown to clarify their meaning. In this configuration, $R_1 = R_2$, and the differential gain is given by the equation

$$A_{v(d)} = 1 + \frac{2R_1}{R_G}$$

Although the resulting IA is not ideal, the results are surprising good.

PROCEDURE
Instrumentation Amplifier Measurements
1.   Measure and record the values of the resistors listed in Table 43–1. In step 3, you will use the measured values of R_1 and R_G for computing the differential gain. For best results, R_1 and R_2 should match.

Table 43–1

Resistor	Listed Value	Measured Value
R_1	10 kΩ	
R_2	10 kΩ	
R_G	470 Ω	
R_3	10 kΩ	
R_4	10 kΩ	
R_5	10 kΩ	
R_6	8.2 kΩ	
R_8	100 kΩ	
R_9	100 kΩ	

2. Construct the circuit shown in Figure 43–2. (For this test, R_9 is shorted by the function generator.) The instrumentation amplifier (IA) is shown in the dotted box. It is driven by the generator in differential mode. The purpose of R_8 and R_9 is to assure a bias path for the op-amps when the source is isolated in step 6. To simplify the schematic, the power supply connections and bypass capacitors are not shown. (Use ± 15 V for all op-amps; use two 1.0 μF bypass capacitors installed near one of the op-amps as in earlier experiments.)

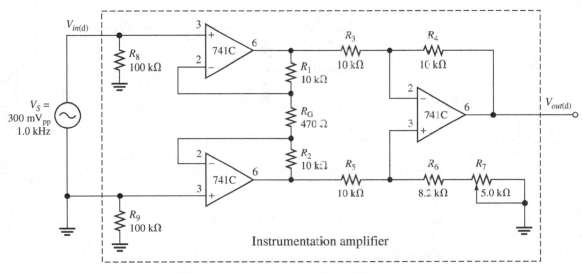

Figure 43–2 Circuit to measure differential parameters.

3. Set the potentiometer (R_7) in the middle of its range. ($R_6 + R_7$ should add to approximately 10 kΩ.) Set the input for a 300 mV$_{pp}$ sine wave at 1.0 kHz. This represents the differential-mode input signal, $V_{in(d)}$. Compute the differential gain from the equation given in the Summary of Theory. Using the computed gain, compute the expected differential output voltage $V_{out(d)}$. Then measure these parameters and record the measured values in Table 43–2.

Table 43–2

Step	Parameter	Computed Value	Measured Value
	Differential input voltage, $V_{in(d)}$	300 mV$_{pp}$	
3	Differential gain, $A_{v(d)}$		
	Differential output voltage, $V_{out(d)}$		
	Common-mode input voltage, $V_{in(cm)}$	10 V$_{pp}$	
4	Common-mode gain, $A_{v(cm)}$		
	Common-mode output voltage, $V_{out(cm)}$		
5	CMRR'		

4. Drive the IA with a common-mode signal as shown in Figure 43–3. Set the signal generator for a 10 V_{pp} signal at 1.0 kHz ($V_{in(cm)}$) and measure this input signal. Observe the common-mode output voltage and adjust R_7 for *minimum* output. Measure the peak-to-peak output voltage, $V_{out(cm)}$. Determine the common-mode gain, $A_{v(cm)}$, by dividing the measured $V_{out(cm)}$ by the measured $V_{in(cm)}$. Record all values in Table 43–2.

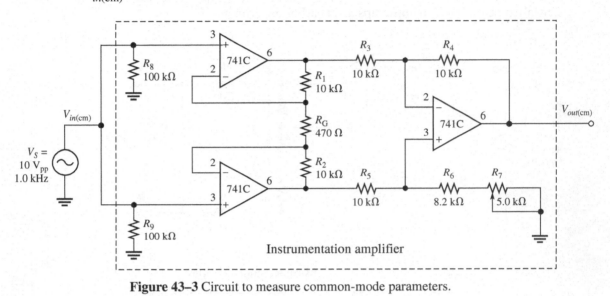

Figure 43–3 Circuit to measure common-mode parameters.

5. Determine the CMRR′ (in dB) from the equation CMRR′ = 20 log ($A_{v(d)}/A_{v(cm)}$). Enter this as the measured value in Table 43–2. When you have completed the common-mode measurements, reduce the signal generator frequency to 60 Hz. This will represent the noise source for the last part of the experiment.

Adding Differential-mode and Common-mode Sources

6. In this step, you will build a pulse oscillator to serve as a differential signal source for the instrumentation amplifier. The oscillator is a 555 timer (a small integrated circuit that is easy to connect as an oscillator). Construct the circuit shown in Figure 43–4 (preferably on a separate protoboard if you have one available). R_C and R_D serve as an output voltage divider to simulate a small signal source (such as a transducer). Measure the output frequency and voltage and indicate these values in the first two rows of Table 43–3.

Note that the differential signal source must be *floating* (no common ground with IA) so it is powered by a small 9 V battery as shown.

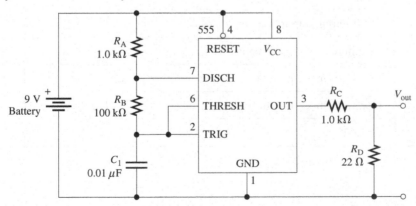

Figure 43–4 Oscillator to serve as a source for the instrumentation amplifier.

354

Table 43–3

Parameter	Measured Value
Oscillator frequency	
$V_{out(pp)}$ from oscillator	
$V_{out(pp)}$ from IA	

7. Connect the oscillator to the IA as shown in Figure 43–5 with about 30 cm of twisted-pair wire to simulate a short transmission line. Be sure there is no common ground from the oscillator to the input of the instrumentation amplifier. Measure the output signal from the IA and record the output in the last row of Table 43–3. Note that this signal represents a differential-mode input and is therefore amplified by the differential gain, $A_{v(d)}$.

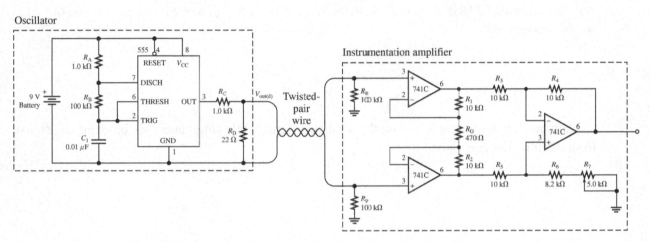

Figure 43–5 Connecting the oscillator to the instrumentation amplifier.

8. Now you will add a simulated source of common-mode noise to the oscillator. Often, the noise source is 60 Hz power line interference, but seldom is it as large or well connected as in this step. Set up your function generator for a 10 V_{pp} sine wave at 60 Hz to simulate a large amount of common-mode interference from a power line. Connect the generator to one side of the oscillator as shown in Figure 43–6. Observe the output signal from the IA. Adjust R_7 for minimum common-mode signal.

Observations: _____

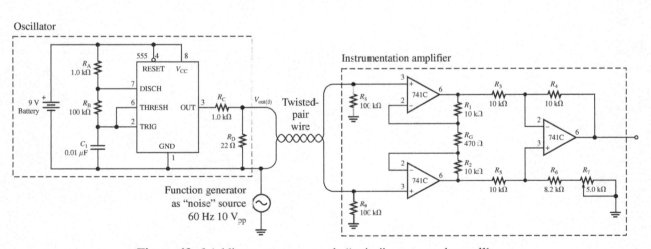

Figure 43–6 Adding a common-mode "noise" source to the oscillator.

CONCLUSION

EVALUATION AND REVIEW QUESTIONS

1. What type of noise could *not* be rejected by an instrumentation amplifier?

2. Some instrumentation amplifiers have a CMRR′ of 130 dB. Assuming the circuit in this experiment had a CMRR′ of 130 dB with the same differential gain, what common-mode output would you expect?

3. Explain how the IA was able to pass the oscillator signal while simultaneously blocking the signal from the function generator.

4. Why was it important to power the 555 timer using a separate battery rather than use the same supply that powered the op-amps?

5. What advantage does an IA, such as the one that you constructed, have over an ordinary differential amplifier?

FOR FURTHER INVESTIGATION

Many transducers convert a physical quantity (light, pressure, speed) into an oscillation. The resulting frequency will depend on a capacitance or resistive change in the transducer. This type of change can easily be observed in this experiment by replacing R_B in the oscillator with a CdS cell. The resistance of a CdS cell is a function of the incident light.

Replace R_B with a CdS cell and observe the effect on the output signal from the IA as you cover the cell. Measure the lowest and highest frequency you can obtain. Can you think of an application for this circuit? Summarize your results.

MULTISIM TROUBLESHOOTING

This experiment has four Multisim files on the website (www.prenhall.com/floyd). Three of the four files contain a simulated "fault"; one has "no fault". The file with no fault is named EXP43-2-nf. You may want to open this file to compare your results with the computer simulation. Then open each of the files with faults. Use the simulated instruments to investigate the circuit and determine the problem. The following are the filenames for circuits with troubleshooting problems for this experiment.

EXP43-2-f1
   Fault:

EXP43-2-f2
   Fault:

EXP43-2-f3
   Fault:

44 Active Diode Circuits

Name _____
Date _____
Class _____

READING
Text, Section 20–4

OBJECTIVES
After performing this experiment, you will be able to:
1. Explain how an active diode circuit differs from a passive diode circuit.
2. Test two active diode clipping circuits, comparing the output and input signals.
3. Test an active diode dc restoring circuit.

MATERIALS NEEDED
Resistors:
> Three 10 kΩ, one 1.0 MΩ

Capacitors:
> One 0.1 μF, two 1 μF

One 10 kΩ potentiometer
Two signal diodes 1N914 (or equivalent)
741C op-amp
For Further Investigation:
> Second 741 C op-amp
> Two additional 1.0 μF capacitors
> Three additional 10 kΩ resistors

SUMMARY OF THEORY
An active diode circuit is one that includes an op-amp to "idealize" the normal diode response. Instead of a 0.7 V offset and the forward resistance of a diode, the active diode has nearly zero offset and an idealized response. Active diode circuits have broad application in signal processing, where a circuit changes an input signal in some way. Included in this category are rectification, clipping, clamping, and various conversions. The "active" part of the diode circuit is an op-amp. For low frequencies (<1 kHz), the dependable 741C works fine. For higher speeds, a FET input amplifier with a faster slew rate is a better choice.

A common active diode circuit is the precision half-wave rectifier shown in Figure 44–1(a). Notice that a diode is inside the feedback loop. When the diode is forward-biased by a positive-going input signal, the feedback loop is complete and the circuit acts as an ordinary voltage-follower with V_{out} following V_{in}. The high gain of the op-amp automatically adjusts the diode's driving voltage until the output replicates the input, even for input signals less than 1 mV. A negative-going input signal causes the op-amp output to reverse-bias the diode, breaking the feedback loop, and disconnecting the output. Without feedback, the op-amp goes into negative saturation, keeping the diode off. In this case, the output is at zero volts.

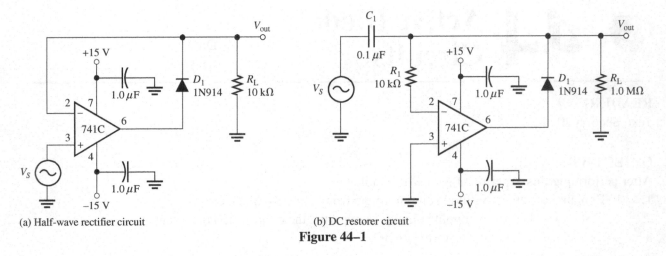

(a) Half-wave rectifier circuit

(b) DC restorer circuit

Figure 44–1

The 741C is limited to frequencies below about 1 kHz for the clipping circuit in this experiment because of slew rate limitations. This occurs when the op-amp output must swing from negative saturation to a positive output as the signal crosses the zero level. Although faster op-amps are available, the 741C will work fine as long as this limitation is not a problem.

In this experiment, you will investigate two clipping circuits. The first clipping circuit is the precision rectifier circuit just described; the second is a modification that allows a variable clipping level, but with the disadvantage of high output impedance. A second op-amp, connected as a voltage-follower, corrects this problem.

A small modification to the second clipping circuit will produce a dc restorer (or clamping) circuit. The dc restorer circuit gets its name from TV circuits, where it restores the original dc level that is lost in capacitively coupled amplifiers. The basic circuit is shown in Figure 44–1(b); it is recognized by a series capacitor with the input signal. The output of this basic circuit is a replica of the input except the dc level is shifted so that the output is always positive in the circuit shown. The capacitor is charged to a dc level that is equal to the input peak signal, which is then added to the input to form the shifted output. Because the output impedance is high, the circuit requires a very high impedance load; notice that R_L is 1.0 MΩ.

After investigating the clipping circuits and dc restorer, you will investigate a peak detector circuit as shown in Figure 44–2. In this circuit, resistor R_1 protects the input from capacitive discharge when the power is removed form the op-amp. Resistor R_2 is included to balance the input bias currents. The circuit works by charging the capacitor to the peak of the input signal level. At the price of slight variation in the output level, the load resistor provides a discharge path for the capacitor to speed up the circuit response time when the input level changes.

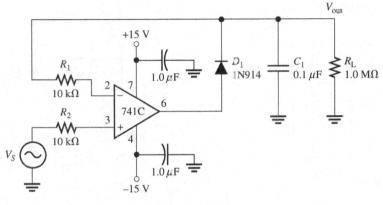

Figure 44–2 Peak detector.

PROCEDURE

Clipping Circuits

1. The resistance values of the six 10 kΩ resistors in this experiment are not critical. You should double check that each resistor is within 10% of 10 kΩ.

2. Construct the active diode circuit half-wave rectifier circuit shown in Figure 44–1. Connect the input (pin 3) to your function generator set to a 4 V_{pp} sine wave at 200 Hz. Verify that the offset voltage from the function generator is 0 V.

3. Sketch the waveform on Plot 44–1. Label the axis and show both the input and output signals.

Plot 44–1

4. To see the effect of slew rate limiting on the output, raise the frequency to 2 kHz and observe the output. You may notice some distortion on the output above 1 kHz. Sketch the signal in Plot 44–2. Temporarily, add a second diode, D_2, as shown in Figure 44–3, which will prevent the output from going into negative saturation. This technique works well to remove the distortion and increase the frequency response from the positive output rectifier. Explain how this change affects the signal at 2 kHz.

_____

_____

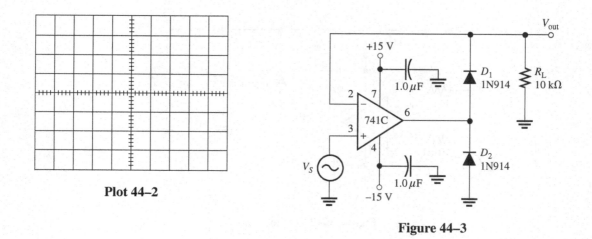

Plot 44–2

Figure 44–3

5.   Remove D_2. Restore the frequency to 200 Hz and reverse diode D_1. Sketch the waveform on Plot 44–3.

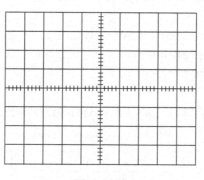

Plot 44–3

6.   The half-wave rectifier circuit you constructed is basically a precision clipping circuit for either positive or negative signals (depending on the orientation of the diode). Figure 44–4 shows a modification of the half-wave circuit to allow a variable clipping level set by V_{REF}. The purpose of R_1 is to isolate the function generator and increase the input impedance. Diode D_1 has been restored to its original orientation. Notice the signal is applied to the inverting side of the op-amp and the load resistor is much larger to avoid distortion. Compare the output and input as you vary R_3. Describe your observations.

_____

_____

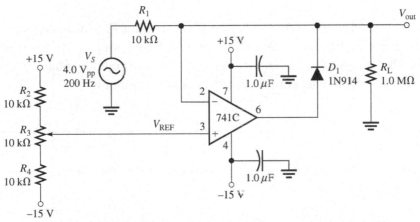

Figure 44–4 A clipping circuit with variable clipping level.

7. Reverse diode D_1 and observe the effect. Describe your observations.

_____

_____

DC Restorer

8. The key to recognizing the dc restorer is to notice the series capacitor with the input. This capacitor is charged to the reference voltage (V_{REF}) set by R_3. The ac portion of the signal is passed directly to the load resistor, which adds to the dc voltage on the capacitor. The purpose of R_1 is to isolate the input from the capacitor when power is removed from the op-amp. Connect the circuit shown in Figure 44–5 and observe V_{out} as you vary R_3. Describe your observations.

_____

_____

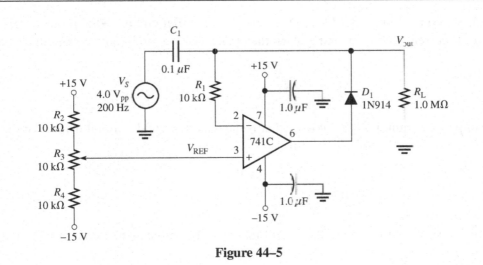

Figure 44–5

9. Reverse diode D_1 and observe the effect on the output as you vary R_3. Describe your observations.

_____

_____

Peak Detector

10.   Construct the peak detector shown in Figure 44–2. Resistor R_1 protects the input of the op-amp in case the capacitor is charged when power is removed. Resistor R_2 provides balance for the bias current. Set the source for a 200 Hz sine wave. Observe the output voltage while you vary the amplitude of the function generator. Describe your observations.

_____

_____

CONCLUSION

EVALUATION AND REVIEW QUESTIONS

1.   For the half-wave rectifier circuit in Figure 44–1(a), assume the function generator input (on pin 3) is replaced with a +1.0 V dc source.
        (a)   What voltage would you expect to measure on pin 6 of the op-amp? _____
        (b)   What voltage would you expect to measure on pin 2 of the op-amp? _____

2.   For the half-wave rectifier circuit in Figure 44–1(a), assume the function generator input (on pin 3) is replaced with a −10 mV dc source.
        (a)   What voltage would you expect to measure on pin 6 of the op-amp? _____
        (b)   What voltage would you expect to measure on pin 2 of the op-amp? _____

3.   In step 4, a second diode was added to the half-wave rectifier circuit. What protects the diode from excess current and overheating when the op-amp tries to go into negative saturation?

4.   For the clipping circuit in Figure 44–4, what is the maximum and minimum value of the reference voltage?

5.   Why was it necessary to increase the size of load resistor on the dc restoring circuit in Figure 44–5?

FOR FURTHER INVESTIGATION

Figure 44–6 shows an interesting circuit that uses two op-amps. The first is an inverting amplifier with a diode in the feedback circuit. (What is the gain?) The second is a basic summing amplifier that adds two inputs together (and inverts the sum). All resistors are the same value. Try to figure out how the circuit will respond to the input. Then build the circuit and write a short summary of what you found. Try varying slightly the input dc offset and notice what happens. What happens if the diode is reversed?

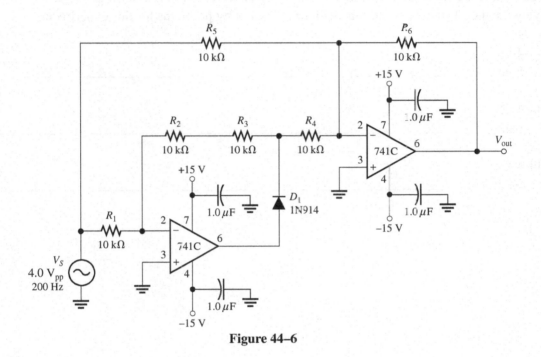

Figure 44–6

MULTISIM TROUBLESHOOTING

This experiment has four Multisim files on the website (www.prenhall.com/floyd). Three of the four files contain a simulated "fault"; one has "no fault". The file with no fault is named EXP44-4-nf. You may want to open this file to compare your results with the computer simulation. Then open each of the files with faults. Use the simulated instruments to investigate the circuit and determine the problem. The following are the filenames for circuits with troubleshooting problems for this experiment.

EXP44-4-f1

Fault: _____

EXP44-4-f2

Fault: _____

EXP44-4-f3

Fault: _____

Application Assignment 20

Name _____

Date _____

Class _____

REFERENCE

Text, Chapter 20; Application Assignment: Putting Your Knowledge to Work

Step 1   Check the Amplifier Board.

     1.   Make sure the circuit board in Figure AA-20-1 is correctly assembled by checking it against the schematic, Figure 20–43 in the text.

     2.   Label the component and input/output designations in agreement with the schematic.

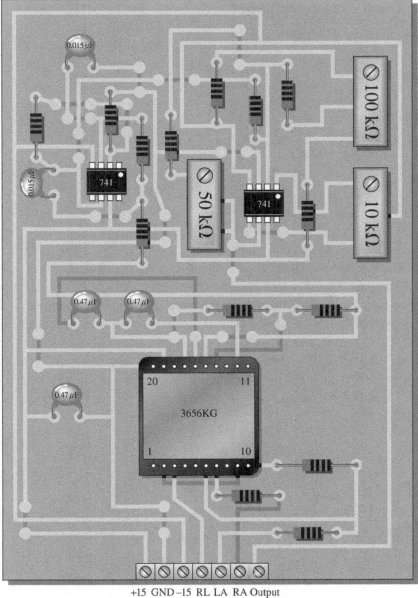

+15 GND –15 RL LA RA Output

Figure AA–20–1

Step 2   Analyze the circuits.

1.   Determine the voltage gain of the isolation amplifier. _____

2.   Determine the bandwidth and the voltage gain of the filter using the formulas given in the text. $BW =$ _____ $A_v =$ _____

3.   Determine the minimum and maximum gains of the postamplifier.

$A_{v(min)} =$ _____ $A_{v(max)} =$ _____

4.   Determine the overall gain range of the amplifier board. _____

5.   Determine the dc voltage range at the wiper of the position adjustment potentiometer. _____

Step 3   Troubleshooting the circuits.

Most likely problem with board 1: _____

Most likely problem with board 2: _____

Most likely problem with board 3: _____

Most likely problem with board 4: _____

RELATED EXPERIMENT

MATERIALS NEEDED

Resistors:
  One 3.3 kΩ, one 5.6 kΩ, two 100 kΩ

Capacitors:
  Two 0.015 μF, two 1.0 μF
  One 741C op-amp

DISCUSSION

The low-pass filter, which is designed to eliminate signals that are not from the heart, can be tested easily as a separate block. Construct the low-pass filter, shown in Figure 20–43 of the text, and measure the cutoff frequency. (The 1.0 μF capacitors, shown on the materials needed list, are not on the schematic but should be inserted to bypass the power supply). The easiest way to measure the cutoff frequency is to put a 2.5 Vpp 20 Hz sine wave from a function generator to the left side of R_6. (Set the input to 500 mV/div to spread the signal over 5 divisions). Then observe the output signal and, using the vernier (fine) voltage control on your scope, make it *appear* to have the same amplitude as the input (you may observe a slight delay in the output). Then raise the frequency until the output is 70.7% of the value at 20 Hz. Report the cutoff frequency.

Checkup 20

REFERENCE

Text, Chapter 20; Lab manual, Experiments 43 and 44

1.   A desirable characteristic of an instrumentation amplifier is
        (a)   high gain for common-mode signals
        (b)   high input impedance
        (c)   both of the above
        (d)   none of the above

2.   In Experiment 43 the differential mode signal from the oscillator was powered by a battery. The main purpose was to:
        (a)   avoid having a common ground
        (b)   enable remote operation of the oscillator
        (c)   prevent common-mode signals from entering the system
        (d)   increase the gain of the circuit

3.   An instrumentation amplifier will tend to block
        (a)   noise introduced in common-mode
        (b)   noise introduced in differential-mode
        (c)   both of the above
        (d)   none of the above

4.   To measure the CMRR′ of an instrumentation amplifier, the gain that must be determined is the
        (a)   common-mode gain
        (b)   differential-mode gain
        (c)   both of the above
        (d)   none of the above

5.   The gain of an OTA is specified in terms of
        (a)   output voltage divided by input current
        (b)   output current divided by input current
        (c)   output current divided by input voltage
        (d)   output voltage divided by input voltage

6.   The gain of an OTA is set by
        (a)   a single input resistor
        (b)   the load resistor
        (c)   the ratio of two resistors in the feedback path
        (d)   the bias current

7. A precision half-wave rectifier is also a
   (a)   full-wave rectifier
   (b)   clamping circuit
   (c)   peak detector
   (d)   precision clipping circuit

8. A circuit that shifts the dc level of a signal is called a
   (a)   clamping circuit
   (b)   dc restorer
   (c)   both of the above
   (d)   none of the above

9. A circuit that has a capacitor in series with the input is a
   (a)   dc restorer
   (b)   clipping circuit
   (c)   peak detector
   (d)   half-wave rectifier

10. To speed up the response of the peak detector to a changing input, you could
    (a)   select an op-amp with a higher slew rate
    (b)   choose a smaller load resistor
    (c)   choose a larger capacitor
    (d)   all of the above

11. What is the purpose of the variable resistor, R_7, in the instrumentation amplifier tested in Experiment 43?

12. R_G controls the gain of the instrumentation amplifier in Figure 43–2. If it is open, what is the gain?

13. To what voltage is C_1 charged to in the circuit of Figure 44–1(b)?

14. The peak detector circuit in Figure 44–2 uses a 1.0 MΩ resistor as a load. Predict the effect if the load is smaller.

15. Predict the effect on the peak detector circuit in Figure 44–2 if the diode is reversed.

45 The SCR

Name _____
Date _____
Class _____

READING
Text, Section 21–6

OBJECTIVES
After performing this experiment, you will be able to:
1. Measure the gate trigger voltage and holding current for an SCR.
2. Compare a transistor latch circuit with an SCR.
3. Explain the effect of varying the gate control on the voltage waveforms in an ac SCR circuit.

MATERIALS NEEDED
Resistors:
  One 160 Ω, two 1.0 kΩ, one 10 kΩ
One 10 kΩ potentiometer
One 0.1 μF capacitor
One LED
One 2N3904 NPN transistor (or equivalent)
One 2N3906 PNP transistor (or equivalent)
One SK3950 SCR (or equivalent)
One 12.6 V power transformer
For Further Investigation:
  One CdS photocell (Jameco 120299 or equivalent)

SUMMARY OF THEORY
Thyristors are a class of semiconductor devices consisting of multiple layers of alternating P and N material. They are bistable devices that use either two, three, or four terminals to control either ac or dc. Thyristors are primarily used in power-control and switching applications. A variety of geometry and gate arrangements are available, leading to various types of thyristors such as the diac, triac, and silicon-controlled rectifier (SCR). In this experiment, you will investigate an SCR. It is one of the oldest and most popular thyristor. It is a four-layer device and can be represented as equivalent PNP and NPN transistors, as shown in Figure 45–1.

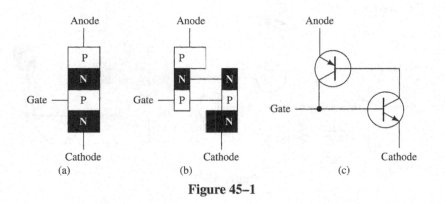

Figure 45–1

The SCR is a thyristor that operates as a latching switch controlled by a sensitive gate. If the anode is negative with respect to the cathode, the SCR is reverse-biased and will be off. When the anode is made more positive than the cathode, the SCR is forward-biased, but without a gate signal, it remains off. The application of a small positive gate pulse causes the SCR to go rapidly into conduction. Once it begins conduction, control is lost by the gate. Conduction ceases only when the anode current is brought below a value called the *holding current.*

SCRs have specific requirements for proper triggering. A number of special thyristors and other solid-state devices, such as the unijunction transistor (UJT), are used for trigger circuits. The primary requirement of any triggering circuit is to provide adequate gate current and voltage at a precise time. The triggering device provides a precise control signal to a thyristor power device. Applications include dc switching, motor control, electronic ignition, battery chargers, and lamp drivers.

PROCEDURE

1. Measure and record the resistance of the resistors listed in Table 45–1. R_2 is a 10 kΩ variable resistor, so it is not listed.

Table 45–1

Resistor	Listed Value	Measured Value
R_1	1.0 kΩ	
R_3	160 Ω	
R_4	1.0 kΩ	
R_5	10 kΩ	

Table 45–2

	Transistor Latch	SCR
V_{AK} (off state)		
V_{AK} (on state)		
$V_{\text{Gate Trigger}}$		
V_{R4}		
$I_{\text{Holding (min)}}$		

2. Construct the transistor latch shown in Figure 45–2. The purpose of C_1 is to prevent noise from triggering the latch. The switch can be made from a piece of wire. Set R_2 for the maximum resistance and close S_1. The LED should be off. Measure the voltage across the latch shown as V_{AK} (off state). (V_{AK} refers to the voltage from the anode to cathode.) Enter the measured voltage in Table 45–2 under Transistor Latch.

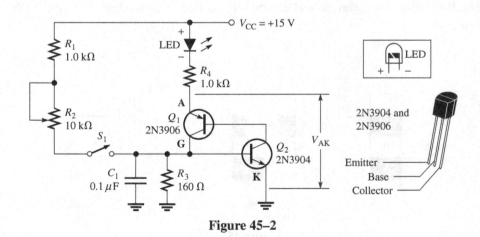

Figure 45–2

372

3.   Slowly decrease the resistance of R_2 until the LED comes on. Measure V_{AK} with the LED on (latch closed). Record this value as V_{AK} (on state) in Table 45–2. Measure the voltage across R_3. Record this as the gate trigger voltage in Table 45–2.

4.   Open S_1. The LED should stay on because of latching action. Connect a voltmeter across R_4. Monitor the voltage while *slowly* decreasing V_{CC}. Record the smallest voltage you can obtain across R_4 with the LED on. Then apply Ohm's law using the measured V_{R4} and the measured resistance of R_4 to compute the current through R_4. This current is the minimum holding current for the latch. Record this as $I_{Holding\ (min)}$ in Table 45–2.

5.   Replace the transistor latch with an SCR as shown in Figure 45–3. Repeat steps 2, 3, and 4 for the SCR. Enter the data in Table 45–2 under SCR.

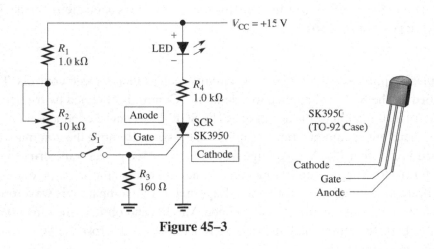

Figure 45–3

6.   As you have seen, the only way to turn off an SCR is to drop the conduction to a value below the holding current. A circuit that can do this for dc operation is shown in Figure 45–4. In this circuit, the capacitor is charged to approximately V_{CC}. When S_2 is momentarily pressed, the capacitor is connected in reverse across the SCR, causing the SCR to drop out of conduction. This is called *capacitor commutation*. Add the commutation circuit to the SCR circuit.

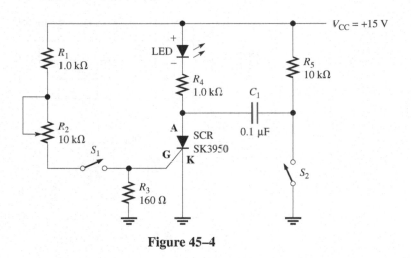

Figure 45–4

7.   Test the commutation circuit by *momentarily* closing S_1 and then *momentarily* closing S_2. Describe your observations.

8.   *Do this procedure only under supervision.*

> **Caution!**   In this procedure, you are instructed to connect a low-voltage (12.6 V ac) transformer to the ac line. Be certain that you are using a properly fused and grounded transformer that has no exposed primary leads. Do not touch any connection in the circuit. At no time will you make a measurement on the primary side of the transformer. Have your connections checked by your instructor before applying power to the circuit.

A typical application of SCRs is in ac circuits such as motor speed controls. The ac voltage is rectified by the SCR and applied to a dc motor. Control is obtained by triggering the gate during the positive alteration of the ac voltage. The SCR drops out of conduction on each negative half-cycle; therefore, a commutation circuit is unnecessary. Remove the commutation circuit and replace V_{CC} with a 12.6 V_{rms} voltage from a low-voltage power transformer.[1] Observe the voltage waveform across R_4 by connecting one channel of your oscilloscope on each side of R_4 and using the difference function as illustrated in Experiment 16. Compare this waveform with the voltage waveform across the SCR anode to cathode. Vary R_2 and observe the effect on the waveforms. On Plot 45–1, sketch representative waveforms across R_4 and across the SCR. Show the measured voltage on your sketch.

V_{R4}

V_{AK}

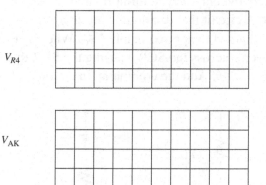

Plot 45–1

---

[1]A signal generator can be used instead. Set the generator for a 15 V peak signal at 60 Hz.

CONCLUSION

EVALUATION AND REVIEW QUESTIONS

1. Explain how to turn off a conducting SCR in a dc circuit.

2. To what does commutation in an SCR circuit refer?

3. Explain why a short from V_{CC} to the anode of the SCR in Figure 45–3 could cause the SCR to burn out.

4. What symptom would you expect to see if the SCR in the circuit of Figure 45–4 had an anode-to-cathode short?

5. For the circuit of Figure 45–4, what effect on the voltage waveform measured across R_4 would you expect if the holding current for the SCR were higher?

FOR FURTHER INVESTIGATION

The trigger control circuit of an SCR can be controlled by a light-sensitive detector such as a photocell. Operate the circuit shown in Figure 45–3 from a sine wave source. Replace R_3 with a CdS photocell. Then put R_3 back and replace R_2 with the photocell. Summarize your findings in a laboratory report. There are various resistive sensors on the market for temperature, moisture, and so forth. Can you think of other potential applications for this circuit?

Application Assignment 21

Name _____

Date _____

Class _____

REFERENCE

Text, Chapter 21; Application Assignment: Putting Your Knowledge to Work

Step 1   Relate the PC board to the schematic. Make sure that the circuit board shown in Figure
AA-21-1 is correctly assembled by checking it against the schematic in text Figure
21–43. Backside connections are shown as darker traces. The motor and the lamp are
external to the board. Label a copy of the board with component and input/output
designations.

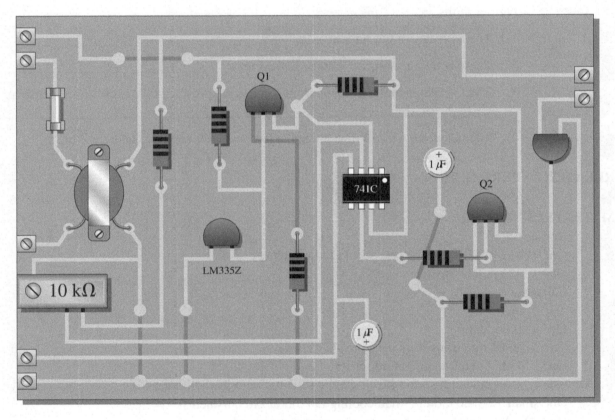

Figure AA-21–1

Step 2   Analyze the circuit. If the LM335Z voltage is 2.93 V at 20°C, determine the voltage at the
inverting input of the comparator. $V_{(-)} =$ _____.
The LM 335Z has breakdown voltage proportional to absolute temperature (K) of
+10 mV/K. Determine the voltage from the LM335Z at +50°C. $V_{(50°C)} =$ _____.

Assume the output of the transformer is 12.6 V rms. To what dc voltage can the noninverting comparator input be adjusted at room temperature (20°C) so that the motor does not run? $V_{(-)} = $ _____ .

Step 3   Test the circuit board. Develop a step-by-step procedure to check the circuit board for proper operation. Specify ac and dc voltage values for all the measurements to be made.

_____

_____

_____

Step 4   Troubleshoot the circuits. For each of the following symptoms, determine the possible fault or faults.

1. The collector of Q2 stays at +15 V. _____

2. There is no pulsating dc voltage to the motor. _____

3. The fan motor runs continuously at a constant speed regardless of the temperature.
    _____

4. The fan motor will not run although it has pulsed dc voltage. The light is on. _____

RELATED EXPERIMENT

MATERIALS NEEDED
Resistors:
One 100 Ω ½ W, one 330 Ω, one 2.7 kΩ, three 10 kΩ
One 10 kΩ potentiometer
Capacitors:
Two 1.0 μF
One 741C op-amp
Two 2N3904 npn transistors (or equivalent)
One LM335Z zener temperature sensor
One type 44 bulb and socket (leads can be soldered onto the bulb instead)
One small 12 V dc motor (small cooling fan motor or equivalent)
One 12.6 V_{rms} power transformer with fused primary side
One SCR type 2N5064 or equivalent (Mouser #519-2N5064)

DISCUSSION
The theory of operation of this circuit is discussed in the text. Construct the circuit and measure the inputs and outputs for the 741C, Q_2, and the SCR. Discuss your observations in a short report.

Checkup 21

Name _____

Date _____

Class _____

REFERENCE

Text, Chapter 21; Lab manual, Experiment 45

1. An inexpensive temperature sensor that is suited for use in a home thermostat is a
    - (a) thermocouple
    - (b) RTD
    - (c) thermistor

2. A temperature sensor that is suited for measuring very high temperatures is a
    - (a) thermocouple
    - (b) RTD
    - (c) thermistor

3. The operating principle of a strain gauge is based on a
    - (a) capacitance change
    - (b) resistance change
    - (c) inductance change
    - (d) temperature change

4. A strain gauge with a higher gauge factor
    - (a) has a higher frequency response
    - (b) has higher resistance
    - (c) is more sensitive
    - (d) all of the above

5. The reference for absolute pressure measurements is
    - (a) atmospheric pressure
    - (b) another pressure
    - (c) a vacuum
    - (d) 1 psi

6. An LVDT is used to measure
    - (a) pressure
    - (b) temperature
    - (c) rotational speed
    - (d) displacement

7. An SCR can be turned off by
    - (a) applying a negative pulse to the gate
    - (b) interrupting the anode supply voltage
    - (c) interrupting the gate current
    - (d) all of the above

8. For an SCR, the *holding current* is the
    (a) current in the gate circuit required to maintain conduction
    (b) minimum current in the gate circuit for triggering
    (c) minimum current in the anode circuit to maintain conduction
    (d) maximum current in the anode circuit to prevent overheating

9. A way to reduce noise when an SCR turns on is to use
    (a) well-regulated power supplies for the anode circuit
    (b) large bypass capacitors in the gate circuit
    (c) commutation circuits to turn off the SCR
    (d) zero voltage switching

10. A device that can be represented as "back-to-back" SCRs is
    (a) two transitors
    (b) a diode
    (c) a triac
    (d) a zero crossing switch

11. Explain what is meant by compensation with respect to thermocouple measurements.

12. How does a three-wire bridge avoid the problem of wire resistance for a remote sensor?

13. Describe two ways to turn off an SCR.

14. When is the best time in the ac cycle to turn on an SCR or triac if avoiding noise is important?

Appendix A:
Project: Constructing a Reed-Switch Motor

READING
Text, Section 7–7

OBJECTIVES
1.   Construct a reed-switch motor.
2.   Test the motor and discuss how you would improve it.

MATERIALS NEEDED
One 1″ × 6″ board approximately 10 inches long
Three popsicle sticks
One nylon round through-hole spacer 3/4″ long, 0.14″ diameter hole
   (Keystone part #888; website is www.keyelco.com)
One nylon washer for #6 screw (Keystone part #3054)
One 1″ long #6 screw, round head
One reed switch (George Risk Industries model PS-2020 3/8″ diameter enclosed
   switch; website is www.grisk.com)
#26 AWG magnet wire (Mouser #501-MW26H)
One 1½ inch long 1/4 inch bolt
Small compass
Masking tape
Two ceramic disk magnets 0.47″ diameter x 0.2″ available from All Magnetics,
   Inc, part number CD14N (website is www.allmagnetics.com)
Sandpaper #60 or #80 grit, 1/4 sheet
Two pieces of 1/2″ stiff foam—about 1″ × 1/2″ for supporting coil and reed switch
Two AA batteries and battery holder

Tools needed: Drill with 1/4 inch bit, hot glue gun, soldering iron, small square, small fine tooth "hobby" saw

SUMMARY OF THEORY
Motors exploit the forces between magnets to turn electrical energy into motion. Electromagnets can be controlled, so they are used in some way in almost all motors.

In this experiment, you will make an electromagnet that provides the field coil for a motor. The field coil will be pulsed by the closure of the reed switch. The rotating armature is constructed from simple materials and uses two permanent magnets to interact with the pulsing electromagnet providing the push to turn the motor.

In constructing the motor, you will work with hot glue. Use caution when you work with hot glue. Do not touch the glue until it cools (a few minutes).

PROCEDURE

1. In this step, you will wind an electromagnet and in the following step, you will use a compass to determine which is the north pole and the south pole of your magnet. The magnet will serve as the field magnet for the motor.

    Obtain a $1\frac{1}{2}$ inch long 1/4 inch bolt and #26 insulated magnet wire. (Magnet wire has a thin varnish insulation). Leave a 6 inch long lead and start wrapping the wire tightly along the bolt. All wraps should go in the same direction. When you have 300 windings, loop a knot in the last winding and again leave another 6 inch long lead. Remove the varnish insulation from the last 1/2 inch of each lead with sandpaper; rotate the wire while sanding to remove the insulation from all sides.

2. Do this step quickly, so as not to drain the battery. Bring a compass near the electromagnet. Touch the ends of your electromagnet across two $1\frac{1}{2}$ V AA batteries in a battery holder. Determine the north end of the electromagnet (recall that like poles repel; the north *seeking* pole of a compass is actually a south pole.) If the head of the bolt is the north pole, note which wire is the positive lead on the electromagnet and mark it with a piece of masking tape. If the head of the bolt is the south end, reverse the connections to the battery and check again. Then mark the positive lead that makes the head of the bolt a north pole.

    Save your electromagnet for step 9.

3. Using the compass, determine the polarity of two ceramic disk magnets. Mark the *south* end of each with an indelible marker. Save the magnets for step 7.

4. Locate a point 4 inches from the end of your board and drill a shallow 1/4″ hole that is just deep enough to accept the head of a #6 screw. Place a small amount of hot glue in the hole and quickly insert the head of the 1″ long screw into the hole. Check that it is straight up and down with a small square before the glue has a chance to set up. See Figure A-1.

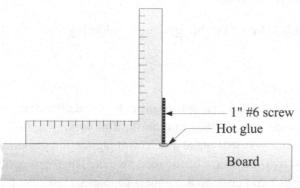

Figure A–1

382

5. Using a small piece of sand paper, square up the ends of two popsicle sticks and sand the sticks so that they are the same length. Mark the center of the sticks with a line and hot glue the 3/4″ nylon spacer to one of the sticks so that it is flush with one side as shown in Figure A-2. It is important that the spacer is centered and perpendicular to the popsicle stick. This will become the rotor for your motor.

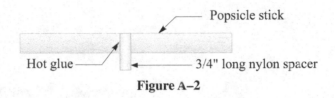

Figure A–2

6. Cut two 1/2″ pieces of popsicle sticks. You may want to use a small hobby saw to cut the popsicle sticks; then lightly sand the ends. Hot glue the 1/2″ pieces to the ends of the rotor to form the assembly shown in Figure A-3.

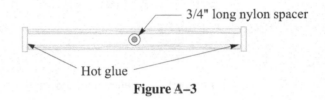

Figure A–3

7. Hot glue the two ceramic magnets from step 3 with the south end to the outside. Place the small nylon washer over the screw and then place the rotor on the screw. Figure A-4 shows the rotor at this stage of the motor project. The rotor should spin easily if you give it a small push.

Figure A–4   Rotor assembly.

8. Hot glue the reed switch to a small piece of 1/2″ foam spacer. Glue the assembly on the board just outside the radius of the rotor as shown in Figure A-5. Hot glue a small 1/2″ foam spacer to the electromagnet, but do not glue it to the board yet.

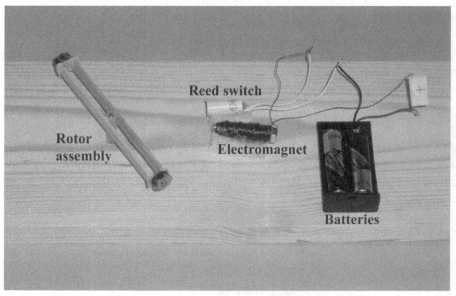

Figure A–5   Final motor assembly.

9. Wire the positive lead on the electromagnet (marked in step 2) to the positive lead of the batteries. Wire the negative lead from the batteries in series with the reed switch as shown in Figure A-6. The electromagnet is positioned close to the reed switch with the head of the bolt (north) pointing away from the rotor. Spin the rotor; you can adjust the location of the electromagnet for best performance. Then glue the electromagnet and the battery pack in place and the motor is completed.

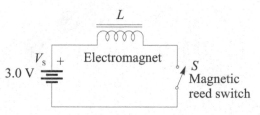

Figure A–6   Schematic of motor.

OBSERVATIONS

QUESTIONS

1. Does the motor work because of attraction between magnets or repulsion? Explain your answer.

2. Why doesn't the motor turn in either direction?

3. Describe with a diagram how you would improve the motor so that it had more speed and power.

SUGGESTED APPLICATION FOR YOUR MOTOR

Look up *Zoetrope* on the Internet. You will find various resources for constructing your own Zoetrope. You can make a Zoetrope from a strip of heavy paper and mount it to the rotor.

Appendix B:
List of Materials for the Experiments

Materials for the Basic Experiments:
(excluding *For Further Investigation* and *Related Experiments*)
Quantities other than one are indicated in parenthesis.

Capacitors:
All capacitors are $\geq$ 35 WV.

1000 pF (4)	0.1 µF (3)
2200 pF	1.0 µF (4)
0.01 µF (4)	10 µF
0.033 µF	47 µF
0.047 µF	100 µF (2)

Diodes:
1N914 signal (2)
1N4001 rectifier (4)
LEDs one red, one green

Inductors:
2 µH (made by winding on a pencil)
25 µH
100 µH
100 nH (2)
$\cong$7 H (secondary of small transformer may substitute)

Integrated Circuits:
555 timer
741C op-amp (3)

Miscellaneous:
Ammeter 0-10 mA
Battery 9 V
Crystal 1.0 MHz
Metric Ruler
Neon bulb (NE-2 or equivalent)
Relay DPDT, 6 V to 12 V dc.
Small speaker (4 Ω to 8 Ω)
Switch SPST
Transformer 12.6 V center tapped
Transformer—small impedance matching

Resistors:
All resistors 1/4 W

22 Ω	2.2 kΩ (2)
47 Ω (2)	2.7 kΩ
68 Ω	4.7 kΩ
100 Ω (2)	5.1 kΩ
150 Ω	5.6 kΩ
220 Ω	6.8 kΩ
270 Ω	8.2 kΩ (4)
330 Ω (2)	10 kΩ (5)
470 Ω	22 kΩ (3)
560 Ω	33 kΩ
680 Ω	47 kΩ (2)
820 Ω	100 kΩ (3)
1.0 kΩ (4)	330 kΩ
1.5 kΩ	470 kΩ
1.6 kΩ	1.0 MΩ
1.8 kΩ	10 MΩ

Transistors and Thyristors
2N3904 small signal npn (3)
2N5458 small signal JFET
SK3950 SCR

Additional Materials for the *For Further Investigation* and Application Assignment *Related Experiments*:
Bulb—one type 1869 or 327 and one type 44
CdS cell Jameco 120299 (or equivalent)
Decade Resistance boxes (2)
Meter calibrator
Regulators:
  7809 or 78L09
  7812 or 78L12
Transistor curve tracer
Variable capacitor 12-100 pF (Mouser ME242-3610-100)
Wheatstone bridge
7414 Schmitt trigger hex inverter
2N5064 SCR (Mouser #519-2N5064 or equivalent)

Appendix C: Manufacturers' Data Sheets

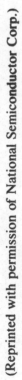

National Semiconductor

Operational Amplifiers/Buffers

LM741/LM741A/LM741C/LM741E Operational Amplifier

General Description

The LM741 series are general purpose operational amplifiers which feature improved performance over industry standards like the LM709. They are direct, plug-in replacements for the 709C, LM201, MC1439 and 748 in most applications.

The amplifiers offer many features which make their application nearly foolproof: overload pro-

tection on the input and output, no latch-up when the common mode range is exceeded, as well as freedom from oscillations.

The LM741C/LM741E are identical to the LM741/LM741A except that the LM741C/LM741E have their performance guaranteed over a 0°C to +70°C temperature range, instead of −55°C to +125°C.

Absolute Maximum Ratings

	LM741A	LM741E	LM741	LM741C
Supply Voltage	±22V	±22V	±22V	±18V
Power Dissipation (Note 1)	500 mW	500 mW	500 mW	500 mW
Differential Input Voltage	±30V	±30V	±30V	±30V
Input Voltage (Note 2)	±15V	±15V	±15V	±15V
Output Short Circuit Duration	Indefinite	Indefinite	Indefinite	Indefinite
Operating Temperature Range	−55°C to +125°C	0°C to +70°C	−55°C to +125°C	0°C to +70°C
Storage Temperature Range	−65°C to +150°C	−65°C to +150°C	−65°C to +150°C	−65°C to +150°C
Lead Temperature (Soldering, 10 seconds)	300°C	300°C	300°C	300°C

Electrical Characteristics (Note 3)

PARAMETER	CONDITIONS	LM741A/LM741E MIN	LM741A/LM741E TYP	LM741A/LM741E MAX	LM741 MIN	LM741 TYP	LM741 MAX	LM741C MIN	LM741C TYP	LM741C MAX	UNITS
Input Offset Voltage	TA = 25°C, RS ≤ 10 kΩ		0.8	3.0		1.0	5.0		2.0	6.0	mV
	RS ≤ 50Ω										mV
	TAMIN ≤ TA ≤ TAMAX			4.0			6.0			7.5	mV
	RS ≤ 50Ω										mV
	RS ≤ 10 kΩ										mV
Average Input Offset Voltage Drift				15							μV/°C
Input Offset Voltage Adjustment Range	TA = 25°C, VS = ±20V	10				±15			±15		mV
Input Offset Current	TA = 25°C		3.0	30		20	200		20	200	nA
	TAMIN ≤ TA ≤ TAMAX			70		85	500			300	nA
Average Input Offset Current Drift				0.5							nA/°C
Input Bias Current	TA = 25°C		30	80		80	500		80	500	nA
	TAMIN ≤ TA ≤ TAMAX			0.210			1.5			0.8	μA
Input Resistance	TA = 25°C, VS = ±20V	1.0	6.0		0.3	2.0		0.3	2.0		MΩ
	TAMIN ≤ TA ≤ TAMAX, VS = ±20V	0.5									MΩ
Input Voltage Range	TA = 25°C										V
	TAMIN ≤ TA ≤ TAMAX	±16			−12	−13		−12	−13		V
		±15						±10	±13		V
Large Signal Voltage Gain	TA = 25°C, RL ≥ 2 kΩ	50			50	200		20	200		V/mV
	VS = ±20V, VO = ±15V										V/mV
	VS = ±15V, VO = ±10V										V/mV
	TAMIN ≤ TA ≤ TAMAX, RL ≥ 2 kΩ	32			25			15			V/mV
	VS = ±20V, VO = ±15V										V/mV
Output Voltage Swing	VS = ±20V	10									V
	RL ≥ 10 kΩ										V
	RL ≥ 2 kΩ										V
	VS = ±15V				±12	±14		±12	±14		V
	RL ≥ 10 kΩ				±10	±13		±10	±13		V
	RL ≥ 2 kΩ										V
Output Short Circuit Current	TA = 25°C	10	25	35		25			25		mA
	TAMIN ≤ TA ≤ TAMAX	10		40							mA
Common-Mode Rejection Ratio	RS ≤ 10 kΩ, VCM = −12V	80	95		70	90		70	90		dB
	RS ≤ 50 kΩ, VCM = −12V										dB

Schematic and Connection Diagrams (Top Views)

Metal Can Package

Order Number LM741H, LM741AH, LM741CH or LM741EH
See NS Package H08C

Dual-In-Line Package

Order Number LM741CN or LM741EN
See NS Package N08B
Order Number LM741CJ
See NS Package J08A

Dual-In-Line Package

Order Number LM741CN-14
See NS Package N14A
Order Number LM741J-14, LM741AJ-14
or LM741CJ-14
See NS Package J14A

(Reprinted with permission of National Semiconductor Corp.)

390

MOTOROLA SEMICONDUCTORS

P.O. BOX 20912 • PHOENIX, ARIZONA 85036

2N3903
2N3904

NPN SILICON ANNULAR TRANSISTORS

. . . designed for general purpose switching and amplifier applications and for complementary circuitry with types 2N3905 and 2N3906.

- High Voltage Ratings — $V_{(BR)CEO}$ = 40 Volts (Min)
- Current Gain Specified from 100 μA to 100 mA
- Complete Switching and Amplifier Specifications
- Low Capacitance — C_{ob} = 4.0 pF (Max)

NPN SILICON
SWITCHING & AMPLIFIER
TRANSISTORS

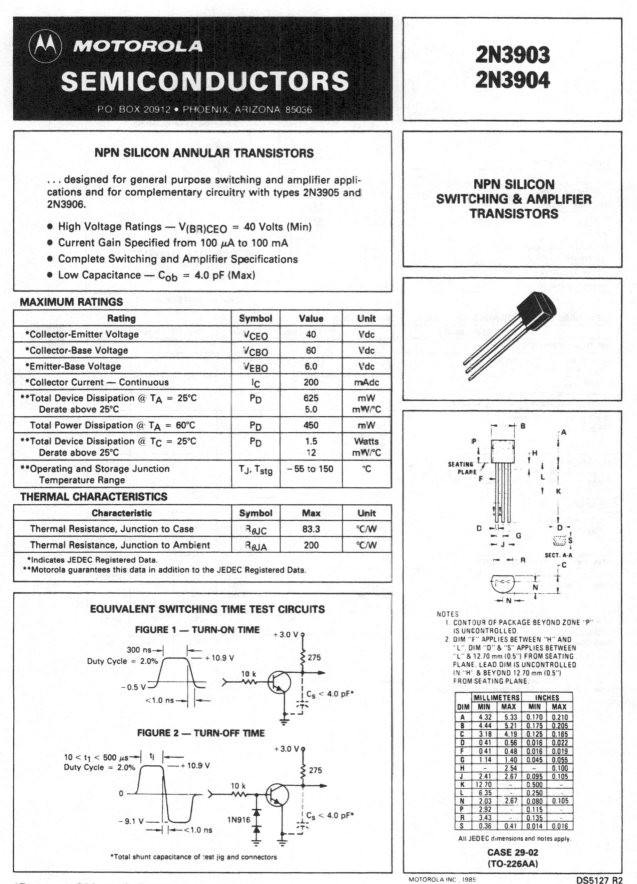

MAXIMUM RATINGS

Rating	Symbol	Value	Unit
*Collector-Emitter Voltage	V_{CEO}	40	Vdc
*Collector-Base Voltage	V_{CBO}	60	Vdc
*Emitter-Base Voltage	V_{EBO}	6.0	Vdc
*Collector Current — Continuous	I_C	200	mAdc
**Total Device Dissipation @ T_A = 25°C Derate above 25°C	P_D	625 5.0	mW mW/°C
Total Power Dissipation @ T_A = 60°C	P_D	450	mW
**Total Device Dissipation @ T_C = 25°C Derate above 25°C	P_D	1.5 12	Watts mW/°C
**Operating and Storage Junction Temperature Range	T_J, T_{stg}	−55 to 150	°C

THERMAL CHARACTERISTICS

Characteristic	Symbol	Max	Unit
Thermal Resistance, Junction to Case	$R_{\theta JC}$	83.3	°C/W
Thermal Resistance, Junction to Ambient	$R_{\theta JA}$	200	°C/W

*Indicates JEDEC Registered Data.
**Motorola guarantees this data in addition to the JEDEC Registered Data.

EQUIVALENT SWITCHING TIME TEST CIRCUITS

FIGURE 1 — TURN-ON TIME

300 ns
Duty Cycle = 2.0%
+10.9 V
−0.5 V
<1.0 ns

+3.0 V
275
10 k
C_S < 4.0 pF*

FIGURE 2 — TURN-OFF TIME

10 < t_1 < 500 μs
Duty Cycle = 2.0%
+10.9 V
0
−9.1 V
<1.0 ns

+3.0 V
275
10 k
1N916
C_S < 4.0 pF*

*Total shunt capacitance of test jig and connectors

NOTES
1. CONTOUR OF PACKAGE BEYOND ZONE "P" IS UNCONTROLLED.
2. DIM "F" APPLIES BETWEEN "H" AND "L". DIM "D" & "S" APPLIES BETWEEN "L" & 12.70 mm (0.5") FROM SEATING PLANE. LEAD DIM IS UNCONTROLLED IN "H" & BEYOND 12.70 mm (0.5") FROM SEATING PLANE.

DIM	MILLIMETERS MIN	MILLIMETERS MAX	INCHES MIN	INCHES MAX
A	4.32	5.33	0.170	0.210
B	4.44	5.21	0.175	0.205
C	3.18	4.19	0.125	0.165
D	0.41	0.56	0.016	0.022
F	0.41	0.48	0.016	0.019
G	1.14	1.40	0.045	0.055
H	–	2.54	–	0.100
J	2.41	2.67	0.095	0.105
K	12.70	–	0.500	–
L	6.35	–	0.250	–
N	2.03	2.67	0.080	0.105
P	2.92	.	0.115	-
R	3.43	–	0.135	-
S	0.36	0.41	0.014	0.016

All JEDEC dimensions and notes apply.

CASE 29-02
(TO-226AA)

MOTOROLA INC . 1985

DS5127 R2

(Courtesy of Motorola Inc.)

CASE 29-04, STYLE 5
TO-92 (TO-226AA)

2 Source

3 Gate

1 Drain

JFET
GENERAL PURPOSE

N-CHANNEL — DEPLETION

Refer to 2N4220 for graphs.

MAXIMUM RATINGS

Rating	Symbol	Value	Unit
Drain-Source Voltage	V_{DS}	25	Vdc
Drain-Gate Voltage	V_{DG}	25	Vdc
Reverse Gate-Source Voltage	V_{GSR}	−25	Vdc
Gate Current	I_G	10	mAdc
Total Device Dissipation @ T_A = 25°C Derate above 25°C	P_D	310 2.82	mW mW/°C
Junction Temperature Range	T_J	125	°C
Storage Channel Temperature Range	T_{stg}	−65 to +150	°C

ELECTRICAL CHARACTERISTICS (T_A = 25°C unless otherwise noted.)

Characteristic		Symbol	Min	Typ	Max	Unit
OFF CHARACTERISTICS						
Gate-Source Breakdown Voltage (I_G = −10 μAdc, V_{DS} = 0)		$V_{(BR)GSS}$	−25	—	—	Vdc
Gate Reverse Current (V_{GS} = −15 Vdc, V_{DS} = 0) (V_{GS} = −15 Vdc, V_{DS} = 0, T_A = 100°C)		I_{GSS}	— —	— —	−1.0 −200	nAdc
Gate Source Cutoff Voltage (V_{DS} = 15 Vdc, I_D = 10 nAdc)	2N5457 2N5458 2N5459	$V_{GS(off)}$	−0.5 −1.0 −2.0	— — —	−6.0 −7.0 −8.0	Vdc
Gate Source Voltage (V_{DS} = 15 Vdc, I_D = 100 μAdc) (V_{DS} = 15 Vdc, I_D = 200 μAdc) (V_{DS} = 15 Vdc, I_D = 400 μAdc)	2N5457 2N5458 2N5459	V_{GS}	— — —	−2.5 −3.5 −4.5	— — —	Vdc
ON CHARACTERISTICS						
Zero-Gate-Voltage Drain Current* (V_{DS} = 15 Vdc, V_{GS} = 0)	2N5457 2N5458 2N5459	I_{DSS}	1.0 2.0 4.0	3.0 6.0 9.0	5.0 9.0 16	mAdc
SMALL-SIGNAL CHARACTERISTICS						
Forward Transfer Admittance Common Source* (V_{DS} = 15 Vdc, V_{GS} = 0, f = 1.0 kHz)	2N5457 2N5458 2N5459	$|y_{fs}|$	1000 1500 2000	— — —	5000 5500 6000	μmhos
Output Admittance Common Source* (V_{DS} = 15 Vdc, V_{GS} = 0, f = 1.0 kHz)		$|y_{os}|$	—	10	50	μmhos
Input Capacitance (V_{DS} = 15 Vdc, V_{GS} = 0, f = 1.0 MHz)		C_{iss}	—	4.5	7.0	pF
Reverse Transfer Capacitance (V_{DS} = 15 Vdc, V_{GS} = 0, f = 1.0 MHz)		C_{rss}	—	1.5	3.0	pF

*Pulse Test: Pulse Width ≤ 630 ms; Duty Cycle ≤ 10%.

6